Laser Processing and Multi-Energy Field Manufacturing of High-Performance Materials

Laser Processing and Multi-Energy Field Manufacturing of High-Performance Materials

Editors

Xiaoxiao Chen
Yaou Zhang
Anhai Li

Basel • Beijing • Wuhan • Barcelona • Belgrade • Novi Sad • Cluj • Manchester

Editors
Xiaoxiao Chen
Ningbo Institute of Materials
Technology and Engineering,
Chinese Academy of Sciences
Ningbo, China

Yaou Zhang
Shanghai Jiao Tong
University
Shanghai, China

Anhai Li
Shandong University
Jinan, China

Editorial Office
MDPI
St. Alban-Anlage 66
4052 Basel, Switzerland

This is a reprint of articles from the Special Issue published online in the open access journal *Materials* (ISSN 1996-1944) (available at: https://www.mdpi.com/journal/materials/special_issues/5WZ2C404JE).

For citation purposes, cite each article independently as indicated on the article page online and as indicated below:

Lastname, A.A.; Lastname, B.B. Article Title. *Journal Name* **Year**, *Volume Number*, Page Range.

ISBN 978-3-0365-8832-2 (Hbk)
ISBN 978-3-0365-8833-9 (PDF)
doi.org/10.3390/books978-3-0365-8833-9

Contents

About the Editors

Xiaoxiao Chen

Dr Xiaoxiao Chen is currently a Professor of the Ningbo Institute of Materials Technology and Engineering, Chinese Academy of Sciences. His main research fields include efficient and precise laser processing theory and technology, digital manufacturing, rotary ultrasonic machining, multi-axis high-speed milling, and multi-energy field composite manufacturing theory, processes, and systems. He has published over 70 papers, and applied for 25 patents and software copyrights. He is the special reviewer for more than 20 international journals. He is a senior member of the Chinese Society of Mechanical Engineering, the Chinese Society of Optics, and the Chinese Society of Aeronautics.

Yaou Zhang

Dr. Yaou Zhang is currently working as an Assistant Professor in the discipline of Mechanical Engineering at the School of Mechanical Engineering, Shanghai Jiao Tong University, Shanghai, China. He received a Ph.D. degree from Shanghai Jiao Tong University in 2008. His areas of interest are electrical discharging machining, multi-energy field processing technology, specialized robots, and smart manufacturing. He served as the principal investigator of numerous competitive research funds. He has published over 50 SCI journal papers and been authorized for over 30 invention patents.

Anhai Li

Dr. Anhai Li is currently an Associate Professor at the School of Mechanical Engineering, Shandong University, Jinan, China. He received his PhD in Mechanical Engineering from Shandong University in 2013. He has been mainly engaged in high-efficiency and precision machining, cutting tools, machined surface integrity, and multi-scale and multi-physics modelling of cutting simulations. From March 2016 to March 2017, he was a Visiting Research Scholar at the Department of Mechanical and Aerospace Engineering, University of Florida. His work has led to more than 60 papers in peer-reviewed journals such as Int J Mech Sci, Mater Des, J Mater Res Technol, Arch Civil Mech Eng, Eng Fail Anal, Int J Adv Manuf Technol, etc., with over 1400 citations. He was one of the winners of the 4th "Higher win excellent mechanical Doctoral Dissertation Award", the outstanding postdoctoral award of Shandong Province, and the Young Scholars Program of Shandong University.....

Preface

With the development of materials science and technology, there is an endless stream of emerging new materials. Those with excellent physical and mechanical properties such as high-temperature resistance, corrosion resistance, and wear resistance can be called high-performance materials, and the processing technology requirements for these materials are endless. The technical demand for material applications is constantly increasing, and advanced materials have been widely used in various fields. At the same time, composite processing technology is also gradually developing. The composite manufacturing of multiple energy fields can benefit from the advantages of various single energies. After the optimization of various energy field combinations, the high-performance processing of materials can be achieved.

The scope of "Laser Processing and Multi-Energy Field Manufacturing of High-Performance Materials" is the processing mechanism, machining quality, material property evolution, and material preparation of lasers and other energy fields. This book summarizes recent advances in the fields of laser processing and multi-energy field composite manufacturing. It covers a variety of topics, including laser cladding, laser coating, laser-based directed energy deposition, laser cutting, laser grooving, laser drilling, electric discharge machining, ultrasonic burnishing, and ultrasonic-vibration-assisted pressing process methods. The effects of lasers, vibrations, electricity and other energies on the properties and processing techniques of various high-performance materials, such as medium-entropy alloys, refractory high-entropy alloys, high-temperature alloy Inconel 718, carbon-fiber-reinforced composites, ceramic-based composites, diamond materials, aluminum alloys, and hard alloys, are fully analyzed and discussed.

The main purpose of this work is to solicit the latest research results in the field of laser and multi-energy field processing, promote the exchange of research in related fields, and promote the development of laser and multi-energy field composite processing. This work also aims to highlight the challenges of processing mechanisms, theories, and technologies, and provide an outlook on future directions.

Xiaoxiao Chen, Yaou Zhang, and Anhai Li
Editors

Editorial

Laser Processing and Multi-Energy Field Manufacturing of High-Performance Materials

Xiaoxiao Chen [1,2,3,*], Yaou Zhang [4] and Anhai Li [5]

1. Research Centre for Laser Extreme Manufacturing, Ningbo Institute of Materials Technology and Engineering, Chinese Academy of Sciences, Ningbo 315201, China
2. Zhejiang Key Laboratory of Aero Engine Extreme Manufacturing Technology, Ningbo 315201, China
3. University of Chinese Academy of Sciences, Beijing 100049, China
4. School of Mechanical Engineering, Shanghai Jiao Tong University, Shanghai 200240, China; yaou_zhang@sjtu.edu.cn
5. School of Mechanical Engineering, Shandong University, Jinan 250061, China; anhaili@sdu.edu.cn
* Correspondence: chenxiaoxiao@nimte.ac.cn

Citation: Chen, X.; Zhang, Y.; Li, A. Laser Processing and Multi-Energy Field Manufacturing of High-Performance Materials. *Materials* **2023**, *16*, 5991. https://doi.org/10.3390/ma16175991

Received: 23 August 2023
Revised: 29 August 2023
Accepted: 30 August 2023
Published: 31 August 2023

The laser is one of the major inventions of the 20th century, along with atomic energy, the computer and semiconductors. Laser processing technology is non-contact, which makes it suitable for the processing and manufacturing of various materials without using cutting forces. During the machining process, the macro-/micro-processing of mechanical motions and the high-speed scanning of galvanometers can be realized. Compared with traditional processing, this has significant advantages in some aspects of the field.

Today, with the development of materials science and technology, there is an endless stream of various new emerging materials. The technical demand for material applications is constantly increasing, and advanced materials have been widely used in various fields. At the same time, composite processing technology is also gradually developing. The composite manufacturing of multiple energy fields can benefit from the advantages of various single energies. After the optimization of various energy field combinations, the high-performance processing of materials can be achieved.

This Special Issue summarizes recent advances in the fields of laser processing and multi-energy field composite manufacturing. The ten articles published in this Special Issue cover a variety of topics, including laser cladding, laser coating, laser-based directed energy deposition, laser cutting, laser grooving, laser drilling, electric discharge machining, ultrasonic burnishing, and ultrasonic-vibration-assisted pressing process methods. The effects of lasers, vibrations, electricity and other energies on the properties and processing techniques of various high-performance materials, such as medium entropy alloys, refractory high entropy alloys, high-temperature alloys Inconel 718, carbon fiber reinforced composites, ceramic based composites, diamond materials, aluminum alloys, and hard alloys, have been fully analyzed and discussed.

This Special Issue aims to showcase the latest achievements in the fields of laser processing and multi-energy field composite manufacturing, solicit the most important discoveries, highlight the challenges of processing mechanisms, theories and technologies, and provide an outlook on future directions. The article publication status of this Special Issue is as follows.

Laser cladding is an effective method for the surface modification of matrix materials. The preparation of functional coatings on metal substrates is an effective method to enhance the surfaces of steel structures, with good serviceability in applications for engineering parts. However, there are few studies on the UB strengthening of MEA laser-cladding coating. Shen et al. [1] investigated the surface properties of two sorts of medium-entropy alloy (MEA) coatings (CoCrNi coating and FeCoNiCr coating) prepared using laser cladding. After cladding, the two prepared coatings were strengthened with ultrasonic burnishing (UB) treatment. Cladding coating samples were comparatively tested before and after being UB-treated in order to investigate the process effects of UB. When compared with corresponding untreated coating samples, the roughness values of the two sorts of UB-treated samples were decreased by 88.7% and 87.6%, the porosities were decreased by 63.8% and 73.4%, and the micro-hardness values were increased by 41.7% and 32.7%, respectively. Furthermore, the two sorts of UB-treated coating samples exhibited better mechanical properties and wear resistance than corresponding untreated samples.

In addition, Xu et al. [2] investigated the effects of the Cr content on the microstructure and properties of the WVTaTiCr$_x$ (x = 0, 0.25, 0.5, 0.75, 1) coating. The trends in the coating's microstructure, hardness, high-temperature oxidation resistance, and corrosion resistance were analyzed by varying the Cr content. As a result, with the increase in Cr, the coating grains were more refined. The coatings were mainly composed of the BCC solid-solution phase, which promotes the precipitation of the Laves phase with the increase in Cr. The addition of Cr greatly improved the hardness, high-temperature oxidation resistance and corrosion resistance of the coating. The WVTaTiCr (Cr$_1$) exhibited superior mechanical properties, especially in terms of its exceptional hardness, high-temperature oxidation resistance and outstanding corrosion resistance. The average hardness of the WVTaTiCr alloy coating reached 627.36 HV. After 50 h of high-temperature oxidation, the oxide weight of WVTaTiCr increased by 5.12 mg/cm^2, and the oxidation rate was 0.1 mg/(cm^2·h). In the 3.5 wt% NaCl solution, the corrosion potential of WVTaTiCr was −0.3198 V and the corrosion rate was 0.161 mm/a.

The profile of the laser beam plays a significant role in determining the heat input on the deposition surface, further affecting the molten pool dynamics during laser-based directed energy deposition. However, there has been little research on the laser–powder interaction and corresponding heat transport of powders under the laser intensity input with super-Gaussian distribution. Chen et al. [3] proposed an improved thermal-fluid model including a laser–powder interaction model and a metal deposition model to explore the thermal-fluid transport and solidification characteristics under two types of laser beams (Gaussian and super-Gaussian) during the single-track L-DED process of Inconel 718. The deposition surface of the molten pool was calculated using the Arbitrary Lagrangian Eulerian moving mesh approach. Several dimensionless numbers were used to explain the underlying physical phenomena under different laser beams. Moreover, the solidification parameters were calculated using the thermal history at the solidification front. It was found that the peak temperature and liquid velocity in the molten pool under the SGB case were lower compared with those for the GB case. Dimensionless numbers analysis indicated that the fluid flow played a more pronounced role in heat transfer compared to conduction, especially in the GB case. The cooling rate was higher for the SGB case, indicating that the grain size could be finer compared with that for the GB case. Finally, the reliability of the numerical simulation was verified by comparing the computed and experimental clad geometry. The work provides a theoretical basis for understanding the thermal behavior and solidification characteristics under different laser input profiles during directed energy deposition.

Primary dendrite arm spacing (PDAS) is a crucial microstructural feature in nickel-based superalloys produced by laser cladding. Currently, most multi-scale simulations of laser cladding use the finite element method with birth-death elements. However, these approaches do not account for changes in the free surface or fluid flow in the molten pool and the latent heat of melting. Jin et al. [4] proposed a multi-scale model that integrates a 3D transient heat and mass transfer model with a quantitative phase-field model to simulate the dendritic growth behavior in the molten pool for laser cladding Inconel 718. The values of the temperature gradient (G) and solidification rate (R) at the S/L interface of the molten pool under different process conditions were obtained via multi-scale simulation and used as the inputs for the quantitative phase field model. The influence of process parameters on the microstructure morphology in the deposition layer was analyzed. The result shows that the dendrite morphology was in good agreement with the experimental result under varying laser power (P) and scanning velocity (V). PDAS was found to be more sensitive to changes in the laser scanning velocity, and as the scanning velocity decreased from 12 mm/s to 4 mm/s, the PDAS increased by 197% when the laser power was 1500 W. Furthermore, smaller PDAS values can be achieved by combining higher scanning velocity with lower laser power.

Laser technology has also played an important role in the field of composite material processing. Xu et al. [5] studied the carbon-fiber-reinforced composite (CFRP) cutting quality. Due to its properties of high specific strength, low density and excellent corrosion resistance, CFRP has been widely used in aerospace and automobile lightweight manufacturing as an important material. A nanosecond laser with a wavelength of 532 nm was applied to cut holes with a 2-mm-thick CFRP plate by using laser rotational cutting technology. The influence of different parameters on the heat-affected zone, the cutting surface roughness and the hole taper was explored, and the cutting process parameters were optimized. Using the optimized cutting parameters, the minimum value of the heat-affected zone, the cutting surface roughness and the hole taper could be obtained; they were 71.7 μm, 2.68 μm and 0.64°, respectively.

Silicon-carbide-fiber-reinforced silicon carbide ceramic matrix composite (CMC-SiC$_f$/SiC) is a typical difficult-to-process material. Chen et al. [6] investigated the grooved morphology of CMC-SiC$_f$/SiC using a dual-beam coupling nanosecond laser. Two kinds of scanning methods were set up according to the relationship between the spatial posture of the dual beams and the direction of the machining path. The CMC-SiC$_f$/SiC grooving experiments were carried out along different feeding directions (transverse scanning and longitudinal scanning) by using a novel dual-beam coupling nanosecond laser. The results showed that the transverse scanning grooving section morphology had a V shape, and the longitudinal scanning groove section morphology had a W shape. The grooving surface depth and width of transverse scanning were larger and smaller than that of longitudinal scanning when the laser parameters were the same. The depth of the transverse grooving was greater than that of the longitudinal grooving when the laser beam was transverse and had longitudinal scanning, and the maximum grooving depth was approximately 145.39 μm when the laser energy density was 76.73 J/cm^2. The thermal conductivity of the fiber had a significant effect on the local characteristics of the grooved morphology when using medium energy density grooving. The obvious recasting layer was produced after the laser was applied to CMC-SiC$_f$/SiC when using high energy density laser grooving.

Water-jet-assisted laser processing technology can introduce new process advantages, such as reducing thermal effects. A novel coaxial-annulus-argon-assisted (CAAA) atmosphere was proposed to enhance the machining capacity of the water-jet-guided laser (WJGL) when dealing with hard-to-process materials in the study by Li et al. [7]. A theoretical model was developed to describe the two-phase flow of argon and the water jet. Simulations and experiments were conducted to analyze the influence of argon pressure on the working length of the WJGL beam, drainage circle size and extreme scribing depth on ceramic matrix composite (CMC) substrates. Single-point percussion drilling experiments were performed on a CMC substrate to evaluate the impact of machining parameters on hole morphology. On these bases, the CAAAWJGL was applied to scribe micro grooves on a CVD diamond, with a large depth-to-width ratio, good consistency and limited defects. The maximum depth-to-width ratio of the groove and depth-to-diameter ratio of the hole reached up to 41.2 and 40.7, respectively. The thorough holes produced by the CAAAWJGL demonstrated superior roundness and minimal thermal damage, such as fiber drawing and delamination. The average tensile strength and fatigue life of the CMCs specimens obtained through CAAAWJGL machining reached 212.6 MPa and 89,463.8 s, exhibiting a higher machining efficiency and better mechanical properties compared to femtosecond and picosecond laser machining. Moreover, groove arrays with a depth-to-width ratio of 11.5, good perpendicularity, and minimal defects on a CVD diamond were fabricated to highlight the feasibility of the proposed machining technology.

In addition, electric discharge machining is also a commonly used special machining method. Electrostatic field-induced electrolyte jet (E-Jet) electric discharge machining

(EDM) is a newly developed micro-machining method. However, the strong coupling of the electrolyte jet liquid electrode and the electrostatic-induced energy prohibited it from being utilized in the conventional EDM process. Zhang et al. [8] proposed a method with two discharge devices connecting in serials to decouple pulse energy from the E-Jet EDM process. Through the automatic breakdown between the E-Jet tip and the auxiliary electrode in the first device, the pulsed discharge between the solid electrode and the solid workpiece in the second device can be generated. With this method, the induced charges on the E-Jet tip can indirectly regulate the discharge between the solid electrodes, giving a new pulse discharge energy generation method for traditional micro EDM. The pulsed variations in current and voltage generated during the discharge process in the conventional EDM process verified the feasibility of this decoupling approach. The influence of the distance between the jet tip and the electrode, as well as the gap between the solid electrode and the workpiece, on the pulsed energy demonstrates that the gap servo control method is applicable. Experiments with single points and grooves indicate the machining ability of this new energy generation method.

Introducing ultrasonic vibration in the manufacturing or material forming process is a very effective innovation. Li et al. [9] investigated the effect of ultrasonic vibration on the fluidity and microstructure of cast aluminum alloys (AlSi9 and AlSi18 alloys) with different solidification characteristics. Furthermore, the impact of ultrasonic vibration on the flow field during the molten metal filling process was analyzed using fluid simulation software (ANSYS-FLUENT). The results show that ultrasonic vibration can affect the fluidity of alloys in both solidification and hydrodynamics aspects. For the AlSi18 alloy without dendrite-growing solidification characteristics, the microstructure was almost not influenced by ultrasonic vibration, and the influence of ultrasonic vibration on its fluidity was mainly in hydrodynamics aspects. That is, appropriate ultrasonic vibration can improve fluidity by reducing the flow resistance of the melt, but when the vibration intensity is high enough to induce turbulence in the melt, the turbulence will increase the flow resistance greatly and decrease fluidity. However, for the AlSi9 alloy, which obviously has dendrite-growing solidification characteristics, ultrasonic vibration can influence solidification by breaking the growing (Al) dendrite, consequently refining the solidification microstructure. Ultrasonic vibration could then improve the fluidity of the AlSi9 alloy, not only from the hydrodynamics aspect but also by breaking the dendrite network in the mushy zone to decrease the flow resistance.

The ultrasonic-vibration-assisted pressing process can improve the fluidity and the uneven distribution of the density and particle size in the WC-Co powder. Chen et al. [10] used three-dimensional spherical models with the aid of the Python secondary development to simulate WC particles with a diameter of 5 μm and Co particles with a diameter of 1.2 μm. The forming process of the powder at the mesoscale was simulated by virtue of the finite element analysis software ABAQUS. The influence of the vibration amplitude on the fluidity, the filling density, and the stress distribution of WC-Co powder when the ultrasonic vibration was applied to the conventional pressing process was investigated. The simulation results show that the ultrasonic vibration amplitude had a great influence on the density of the compact. With an increase in the ultrasonic amplitude, the compact density also increased gradually, and the residual stress in the billet decreased after the compaction. From the experimental results, the size distribution of the billet was more uniform, the elastic after-effect was reduced, and the dimensional instability was improved.

Overall, the works published in this Special Issue show that high performance materials play an important role in the field of advanced manufacturing technology, and various energies such as laser, electric energy, ultrasonic vibration, and fluid energy can exert different technological advantages in the machining process. The introduction of new energy forms or the combination of various energy fields can significantly improve the processing effect for different materials, and this is one of the important development directions of future manufacturing technologies. The editors hope that the readers can discover the advantages of laser and energy field composite processing of high-performance material from the research results of this Special Issue, and gain some inspiration, which will promote the future research work.

Author Contributions: Investigation, writing, editing, supervision, and review, X.C.; supervision and review, Y.Z. and A.L. All authors have read and agreed to the published version of the manuscript.

Acknowledgments: The authors express their gratitude to the *Materials* journal for offering them an academic platform for research where they can contribute and exchange their research on Laser Processing and Multi-Energy Field Manufacturing of High-Performance Materials.

Conflicts of Interest: The authors declare no conflict of interest.

References

1. Shen, X.; Liu, C.; Wang, B.; Zhang, Y.; Su, G.; Li, A. Surface Properties of Medium-Entropy Alloy Coatings Prepared through a Combined Process of Laser Cladding and Ultrasonic Burnishing. *Materials* **2022**, *15*, 5576. [CrossRef] [PubMed]
2. Xu, Z.; Sun, Z.; Li, C.; Wang, Z. Effect of Cr on Microstructure and Properties of WVTaTiCrx Refractory High-Entropy Alloy Laser Cladding. *Materials* **2023**, *16*, 3060. [CrossRef] [PubMed]
3. Chen, B.; Bian, Y.; Li, Z.; Dong, B.; Li, S.; Tian, C.; He, X.; Yu, G. Effect of Laser Beam Profile on Thermal Transfer, Fluid Flow and Solidification Parameters during Laser-Based Directed Energy Deposition of Inconel 718. *Materials* **2023**, *16*, 4221. [CrossRef] [PubMed]
4. Jin, Z.; Kong, X.; Ma, L.; Dong, J.; Li, X. Prediction of Primary Dendrite Arm Spacing of the Inconel 718 Deposition Layer by Laser Cladding Based on a Multi-Scale Simulation. *Materials* **2023**, *16*, 3479. [CrossRef] [PubMed]
5. Xu, J.; Jing, C.; Jiao, J.; Sun, S.; Sheng, L.; Zhang, Y.; Xia, H.; Zeng, K. Experimental Study on Carbon Fiber-Reinforced Composites Cutting with Nanosecond Laser. *Materials* **2022**, *15*, 6686. [CrossRef] [PubMed]
6. Chen, T.; Chen, X.; Zhang, X.; Zhang, H.; Zhang, W.; Liu, G. Study on the Grooved Morphology of CMC-SiC$_f$/SiC by Dual-Beam Coupling Nanosecond Laser. *Materials* **2022**, *15*, 6630. [CrossRef] [PubMed]
7. Li, Y.; Wang, S.; Ding, Y.; Cheng, B.; Xie, W.; Yang, L. Investigation on the Coaxial-Annulus-Argon-Assisted Water-Jet-Guided Laser Machining of Hard-to-Process Materials. *Materials* **2023**, *16*, 5569. [CrossRef] [PubMed]
8. Zhang, Y.; Gao, Q.; Yang, X.; Zheng, Q.; Zhao, W. Research on Electrostatic Field-Induced Discharge Energy in Conventional Micro EDM. *Materials* **2023**, *16*, 3963. [CrossRef] [PubMed]
9. Li, A.; Wang, Z.; Sun, Z. Effect of Ultrasonic Vibration on Microstructure and Fluidity of Aluminum Alloy. *Materials* **2023**, *16*, 4110. [CrossRef] [PubMed]
10. Chen, Y.; Wang, Y.; Huang, L.; Su, B.; Yang, Y. Investigating the Microscopic Mechanism of Ultrasonic-Vibration-Assisted-Pressing of WC-Co Powder by Simulation. *Materials* **2023**, *16*, 5199. [CrossRef] [PubMed]

Article

Surface Properties of Medium-Entropy Alloy Coatings Prepared through a Combined Process of Laser Cladding and Ultrasonic Burnishing

Xuehui Shen [1,2], Chang Liu [1,2], Baolin Wang [1,2,*], Yu Zhang [1,2], Guosheng Su [1,2] and Anhai Li [3]

1   School of Mechanical Engineering, Qilu University of Technology (Shandong Academy of Sciences), Jinan 250353, China
2   Shandong Institute of Mechanical Design and Research, Jinan 250031, China
3   Key Laboratory of High Efficiency and Clean Mechanical Manufacture of MOE, School of Mechanical Engineering, Shandong University, Jinan 250061, China
*   Correspondence: baolinw@qlu.edu.cn

Abstract: The preparation of functional coatings on metal substrates is an effective method to enhance the surface of steel structures with good serviceability in applications for engineering parts. The objective of this research is to analyze the surface properties of two sorts of medium-entropy alloy (MEA) coatings prepared by laser cladding. After cladding, the two prepared coatings were strengthened by ultrasonic burnishing (UB) treatment. Cladding coating samples before and after being UB-treated were comparatively tested in order to investigate the process effects of UB. When compared with corresponding untreated coating samples, the roughness values of the two sorts of UB-treated samples were decreased by 88.7% and 87.6%, the porosities were decreased by 63.8% and 73.4%, and the micro-hardness values were increased by 41.7% and 32.7%, respectively. Furthermore, the two sorts of UB-treated coating samples exhibited better mechanical properties and wear resistance than corresponding untreated samples.

Keywords: medium-entropy alloy; laser cladding; ultrasonic burnishing; surface roughness; wear resistance

Citation: Shen, X.; Liu, C.; Wang, B.; Zhang, Y.; Su, G.; Li, A. Surface Properties of Medium-Entropy Alloy Coatings Prepared through a Combined Process of Laser Cladding and Ultrasonic Burnishing. *Materials* **2022**, *15*, 5576. https://doi.org/10.3390/ma15165576

Academic Editor: Antonio Riveiro

Received: 29 July 2022
Accepted: 11 August 2022
Published: 13 August 2022

Publisher's Note: MDPI stays neutral with regard to jurisdictional claims in published maps and institutional affiliations.

1. Introduction

In recent years, superior to most traditional alloys [1], multi-principle-element alloys (MPEAs) show excellent properties, such as high strength [2], high hardness [3], high thermal stability [4], and excellent corrosion resistance [5]. Hence, they have a great potential to be applied in the fields of energy, aerospace, pipeline engineering, and so on.

A high-entropy alloy is composed of five or over five kinds of elements with equal or nearly equal molar ratios, and the content of each element is between 5 and 35 at.%. As is well known, high-entropy alloys have four main effects, i.e., the lattice distortion effect, cocktail effect, high-entropy effect, and hysteresis diffusion effect [6,7], and thus show significant performance in many aspects. MEAs are derived from high-entropy alloys, composed of three or four elements in accordance with equal or nearly equal molar ratios. Until now, three kinds of single-phase high-/medium-entropy alloys with face-centered cubic (FCC) structures, namely, CoCrNi, FeCoNiCr, and NiCoCrFeMn, have been widely studied [8,9]. Among these popular high-/medium-entropy alloys with FCC structures, CoCrNi has the optimal strength and toughness [10], as well as unique low-temperature fracture toughness surpassing other known materials [11]. As a result, it has become one of the new materials attracting the most attention.

Generally, the preparation of MEA is mainly based on casting and sintering. Moreover, the main elements in MPEAs are usually expensive. These two aspects hinder the practical application of MEAs. Laser cladding is an advanced surface coating preparation technique that has been widely used to modify the surface properties of engineering parts [12,13]. During laser cladding, alloy powders and the metal matrix surface simultaneously melt using a high power density laser beam, and then the liquid metal rapidly solidifies to form a cladding layer metallurgically bonded with the matrix. MEAs can be coated on surfaces of low-cost material substrates, with a surface performance reaching or exceeding that of the whole medium-/high-entropy alloy, which is an effective method to promote the further application of alloys with multiple principal elements.

Recently, some scholars have carried out research on the preparation and property investigation of high-entropy alloy laser cladding coating. Zhang et al. [14] prepared a TiZrNbWMo refractory high-entropy alloy coating by laser cladding on the surface of a 45 steel substrate, and then annealed the sample for 20 h at 800~1200 °C. The results showed that the microstructures of the cladding coating were mainly BCC phase and a small amount of β-TixW1-x precipitated phase. After being annealed, the microstructure of the BCC phase basically remained stable, illustrating that the prepared cladding coating had higher thermal stability. Simultaneously, the proportion of precipitated phase increased, which suggested that the micro-hardness was improved. Liang et al. [15] prepared AlCrFeNi2W0.2Nbx (x = 0.5, 1.0, 1.5, 2.0) cladding coatings on the surfaces of SS304 substrates, and conducted a wear test. The results showed that the principal phase of the cladding coating was the single BCC solid-solution phase. As x was equal to 1.0, 1.5, and 2.0, there appeared a Laves phase, and the micro-hardness of the cladding coating increased with the Nb content increasing.

However, there exists a "rapid heating and sudden solidification" characteristic in the laser cladding process. The substrate material is usually different from that of cladding coating; accordingly, their thermal expansion coefficients are greatly different, resulting in the generation of structural stress and thermal stress. Moreover, the internal pores and defects could cause the uneven distribution of the hardness of the cladding coating, thus affecting the service performance of the cladded components. UB is a kind of surface-strengthening technique to generate a gradient nanostructure surface layer and fine grains [16]. In addition to the significant decrease in surface roughness, UB treatment can induce compressive residual stress in near-surface materials. Thus, by UB treatment, a gradient structure with more refined grains of treated material could be obtained, achieving an improvement in the surface integrity and mechanical properties of the materials [17–19]. At present, UB technology has been widely applied to many materials, such as iron-base alloy, Ti, commercial pure aluminum, Cu, and Ni alloys [20,21]. Previous reports have demonstrated that ultrasonic surface burnishing treatment can produce a nearly polished surface, gradient nanostructured surface layer, and high residual compressive stress on the friction stir-welded joint of a 7075-T651 aluminum alloy. Furthermore, the fatigue property of the joint after treatment was enhanced and the fatigue life was extended by two orders of magnitude [22]. Applying UB technology on a Cu-10Ni alloy, a gradient nanostructured surface layer was obtained with an 80% increase in surface hardness. Additionally, the Cu-10Ni alloy has better corrosion resistance in 3.5wt% solution due to the formation of surface passivation film promoted by the nanoparticle surface during the corrosion process [23]. However, there are few studies on the UB strengthening of MEA laser cladding coating.

Herein, the aim of our study is to improve the wear resistance of a GCr15-bearing steel surface. A laser cladding technique was applied to prepare two kinds of MEA cladding coatings, i.e., CoCrNi coating and FeCoNiCr coating. After cladding, UB strengthening treatment was conducted on the prepared laser cladding coatings. By comparison, the influences of UB treatment on the surface integrity, mechanical, and tribological properties of the cladding coating were studied, and thus some conclusions were drawn.

2. Materials and Methods

2.1. Experiment Materials

The powders of FeCoNiCr and CoCrNi, purchased from Jiangsu VILORY New Material Technology Co., Ltd. (Xuzhou, China) were adopted with a purity of 99.9% and a particle size of 45~105 μm. The GCr15-bearing steel was used as the substrate with a size of Φ 55 mm × 60 mm. The chemical compositions of the FeCoNiCr and CoCrNi powders and substrate are listed in Table 1, respectively. Further, the approximately spherical microstructure of the powders is shown in Figure 1.

Table 1. Chemical compositions of the powders and substrate (At%).

Sample	C	Co	Cr	Mo	Ni	Si	Mn	P	S	Cu	Fe
FeCoNiCr powder	/	Bal.	22.58	/	25.66	/	/	/	/	/	24.60
CoCrNi powder	/	33.95	32.42	/	Bal.	/	/	/	/	/	/
Substrate	0.98	/	1.42	0.05	0.08	0.22	0.25	0.006	0.011	0.03	Bal.

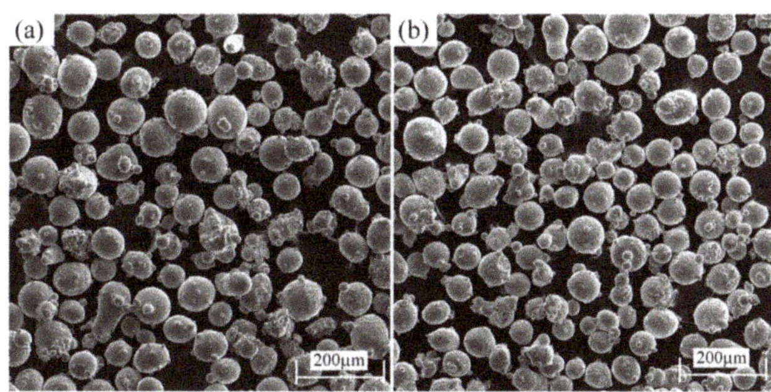

Figure 1. SEM images of MEA powder morphology, (**a**) FeCoNiCr, (**b**) CoCrNi.

2.2. Sample Preparation

In order to prepare the samples, firstly, the GCr15 bearing steel substrate was ground with 180#, 400#, and 600# silicon carbide abrasive papers, respectively, to obtain a smooth surface with the oxide removal. Then, acetone was used to wipe the surface to remove contaminants such as oil stains. Finally, the LYRF-4000W robot laser cladding workstation and LYRF150 high-end cladding integration system were used to prepare the coatings on the substrate. The optimal laser cladding parameters are listed in Table 2, and the processing and photos of laser cladding are illustrated in Figure 2.

Table 2. Laser cladding parameters.

Laser Power (W)	Laser Beam Diameter (mm)	Scanning Speed (mm/s)	Powder Feed Rate (mm/s)	Overlapping Ratio (%)
2000	4	1500	38.4	25

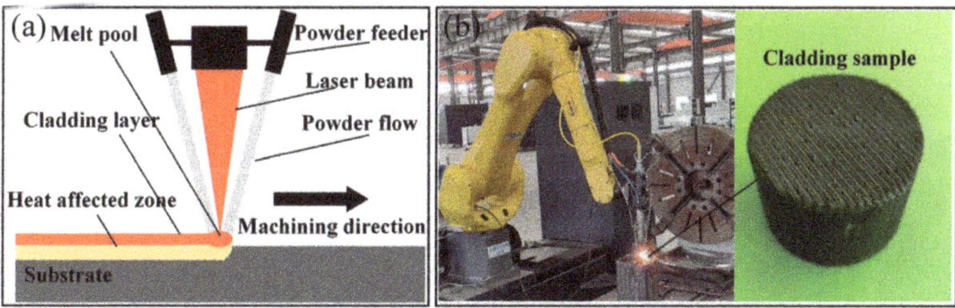

Figure 2. (**a**) Processing of laser cladding, (**b**) photos of laser cladding.

After the laser cladding, the samples were firstly fixed on the CNC lathe for finish turning to obtain a smooth surface. Then, the UB process was conducted on the sample at room temperature. During the experiment process, the static pressure can be obtained by adjusting the extrusion depth of the lathe. The specific parameters are listed in Table 3. The UB processing and experimental setup are presented in Figure 3a,b, respectively.

Table 3. UB parameters.

Frequency (kHz)	Feed Rate (mm/min)	Amplitude (mm)	Spindle Speed (rpm)	Pressure Deep (mm)
28	10	10	160	0.3

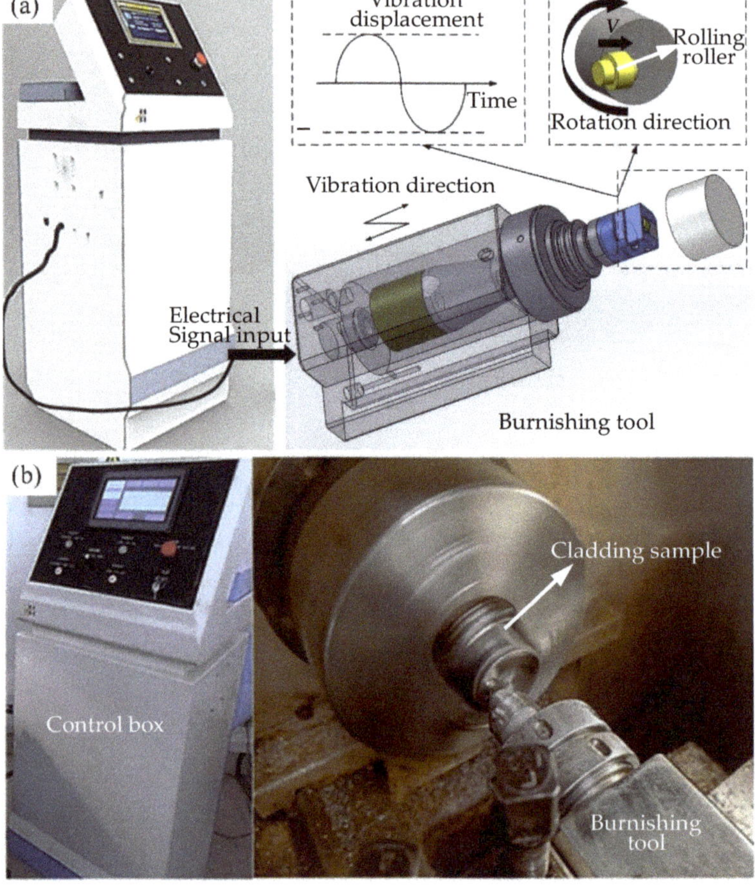

Figure 3. (**a**) Illustration of UB processing, (**b**) photos of the UB experiment setup.

A total of five different samples were prepared for the experiment, as follows. The substrate sample was briefly described as HT(GCr15) as a control sample. Two coating samples, prepared by first laser cladding and then turning finely, were briefly described as LC(FeCoNiCr) and LC(CoCrNi). Furthermore, another two coating samples, prepared by laser cladding, turning finely, and UB treatment, in sequence, were briefly described as UB(FeCoNiCr) and UB(CoCrNi). All samples were cleaned in an ultrasonic cleaner for 10 min after being treated.

2.3. Surface Morphology and Microstructure Characterization

The three-dimensional surface morphologies and roughness of different samples were tested using a white light interferometer with a surface 3D texture topography acquisition system (Contour Elite K, Bruker Company, Karlsruhe, Germany). In order to suppress measurement noise, various spline filters were applied to the data analysis of the surface topography [24–26]. Considering the level of the accuracy of measurement, the averaging method was used. Five different areas of each sample were selected for measuring the surface roughness and taking an average.

The surface microscopic morphologies of the laser cladding coatings of FeCoNiCr and CoCrNi MEAs were observed with an ultra-depth three-dimensional observation microscopic system (VHX-5000, KEYENCE CORPORATION, Osaka, Japan), as well as the microstructures after turning and UB treatment, subsequently, with the analysis for the observation results. The energy dispersive spectrometer (EDS, Phenom ProX, Phenom-World, Eindhoven, The Netherlands) was used to test the chemical composition of the samples. Before the observation, the samples via laser coating and UB treatment were ground with silicon carbide abrasive papers of 180#, 400#, 600#, 1000#, 1500#, and 2000#, respectively. Then, the sample surface was polished to a mirror finish without any scratch with a diamond polishing paste of W0.5. The hydrofluoric acid solution (hydrofluoric acid: nitric acid: water = 2:1:5) was used to etch the polished samples. The porosities of the coatings with different treatments were calculated using Image J software.

2.4. Micro-Hardness and Nano-Indentation Test

A micro Vickers hardness tester (HXD-1000TMC) was used to measure the hardness of the samples with various treatments at a load of 1.961 N ($HV_{0.2}$) for 15 s. The interval between adjacent test points is 100 μm, and each test point was measured three times, taking the average to ensure the data accuracy.

The micro-/nano-mechanical properties of the sample surface were measured and characterized by a nano-indentation tester NHT (MTS, Bruker Company, Karlsruhe Germany) mechanical testing system for microscopic materials. The load of the nano-indentation test was 30 mN, with a loading/unloading speed of 15 mN/min and a load-holding time of 15 s. Moreover, the indentation test was repeated five times for each area to ensure the accuracy and reliability of the data.

2.5. Friction and Wear Test

Friction and wear performance tests were conducted on the substrate sample, as well as those coating samples after turning and UB treatment by an RTEC multifunctional friction and wear tester at room temperature, with a friction way of linear reciprocating motion. Before the test, the sample surface was slightly polished to make the surface smooth without an oxide layer. The counter pair was the GCr15 steel ball with a 7 mm diameter and 60 HRC. The load was 10 N, and the one-way friction distance was 4.5 mm. Furthermore, the reciprocating frequency was 2 Hz, and the friction and wear experiments hold for 15 min. The wear morphology and chemical composition were analyzed and tested by SEM and EDS. Meanwhile, a 3D Super Depth Digital Microscope (VHX-5000 KEYENCE CORPORATION, Osaka, Japan) was used to measure the wear profiles and calculate the wear loss in all testing cases.

3. Results and Discussion

3.1. Surface Roughness

Figure 4 shows the comparison of the surface morphology and surface roughness for the substrate, untreated MEA coating, and UB-treated MEA coating sample surfaces. As can be clearly seen from Figure 4a–c, the turning contour of the substrate and untreated MEA coating samples were relatively stable, but there appeared obvious oscillation corrugations in the turning contour of the laser cladding samples compared with the UB-treated coating sample surfaces, wherein most areas were fattened, as shown in Figure 4d,e. The reason was that during the laser cladding process, the excessively fast cooling velocity resulted in the failure of gas escape in time, uneven residual stress distribution, and defects of void, inclusion, etc., thus affecting the quality of the turning surface. From Figure 4f, there appeared obvious oscillation corrugations on the surface of the laser cladding samples, but the surface roughness parameter Sa (arithmetical average height of surface topography located reference surface [27–29]) of the laser cladding samples slightly decreased. Furthermore, the surface roughness parameter Sa of the UB(FeCoNiCr) and UB(CoCrNi) samples were 52.612 nm and 42.098 nm, respectively, decreasing by 88.7% and 87.6%, compared with the Sa = 465 nm of LC(FeCoNiCr) and the Sa = 340 nm of LC(CoCrNi) samples, which were turned finely only. The reduction of Sa would be beneficial to the improvement of surface properties. The surface roughness of samples via UB treatment became smaller, due to the following two aspects. On the one hand, the rolling head of UB caused the plastic deformation of the sample surface, making the material flow from surface peaks to troughs, thus eliminating turning traces. On the other hand, the rolling head intermittently contacts the sample surface, so the lubrication can fully flow to the working area. This can not only improve the strengthening effect but also reduce the adhesion between the roller and sample surface material. Hence, the surface roughness of the sample by UB treatment significantly decreases [30,31].

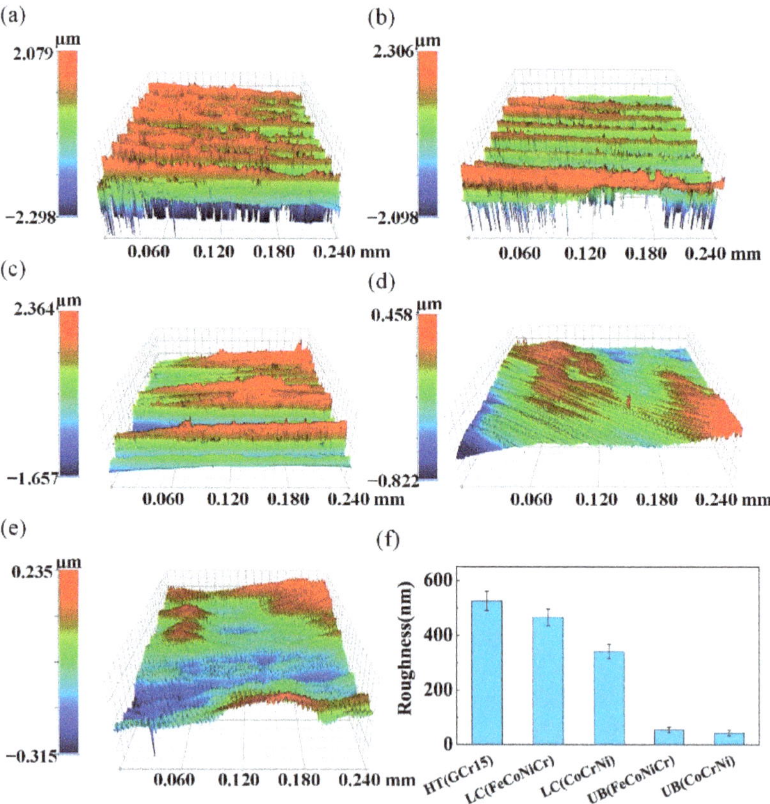

Figure 4. Three-dimensional surface roughness measurements of five samples. (**a**) HT(GCr15), (**b**) LC(FeCoNiCr), (**c**) LC(CoCrNi), (**d**) UB(FeCoNiCr), and (**e**) UB(CoCrNi). (**f**) Comparison of surface roughness parameter Sa of five samples.

3.2. Porosity

With the rapid heating and cooling characteristics, dense and fine grains would be generated in the laser cladding coating. However, due to the excessively rapid cooling speed, the pores inside the laser cladding layer are always present in the laser cladding. To reduce the pores of various cladding coatings, many efforts have been made, especially in improving the laser process parameters [32].

The porosity not only affects the surface roughness but also reduces the mechanical properties of the coating surface, easily leading to high wear rates [33]. Therefore, it is an essential factor to evaluate the surface quality of laser cladding coatings.

Figure 5a–d shows the SEM images of the surface morphology of the four coating samples. As shown in Figure 5a–d, the numbers of pores in Figure 5c,d are significantly less than those in Figure 5a,b. Figure 5e shows the calculated porosities of the four coating samples. As exemplified in Figure 5e, compared with LC(FeCoNiCr) and LC(CoCrNi) coating samples, the surface porosities of the UB-treated coating samples decreased by 63.8% and 73.4%, respectively. Obviously, the UB treatment had a significant effect on decreasing the porosity of the laser cladding coatings. This was because, during the UB process, the external force caused the plastic deformation of the coating sample surface and then the plastic flow in the local deformation area. Accordingly, the pores and defects on the surface were filled, so that the porosity decreased, and then a smooth surface was formed.

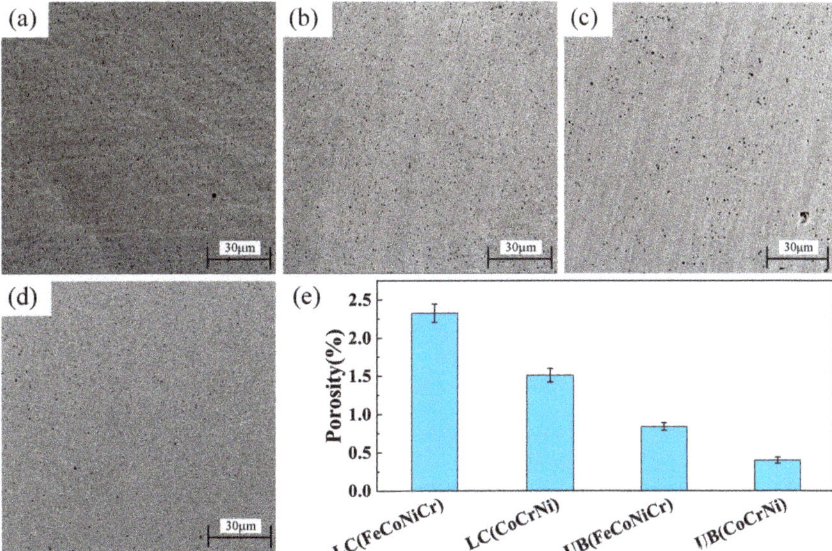

Figure 5. SEM images of coating sample surface morphologies. (**a**) LC(FeCoNiCr), (**b**) LC(CoCrNi), (**c**) UB(FeCoNiCr), (**d**) UB(CoCrNi), and (**e**) the porosities of coating samples.

3.3. Microstructure Characterization

Laser cladding can enable the metallurgical bonding between alloy powders and the metal substrate surface. Owing to the high temperature and element concentration difference, the element diffusion between the laser cladding coating and substrate could change the coating composition and microstructure, thus significantly affecting the coating surface properties. The EDS was used to characterize the chemical composition distribution of MEAs FeCoNiCr and CoCrNi coatings prepared by laser cladding. It was found from the characterization results in Figure 6 that the elements of Fe, Co, Ni, and Cr were evenly distributed in the coating, and there was an obvious transition with the substrate, illustrating the occurrence of metallurgical bonding between the coating and the substrate. Preparing the coating, the substrate had a relatively large dilution effect on the MEA due to the stirring of the molten pool. In the affected layer, the Fe content significantly increased, and near the surface, the element content was basically stable, but the Fe content still increased to a certain extent due to the dilution [34].

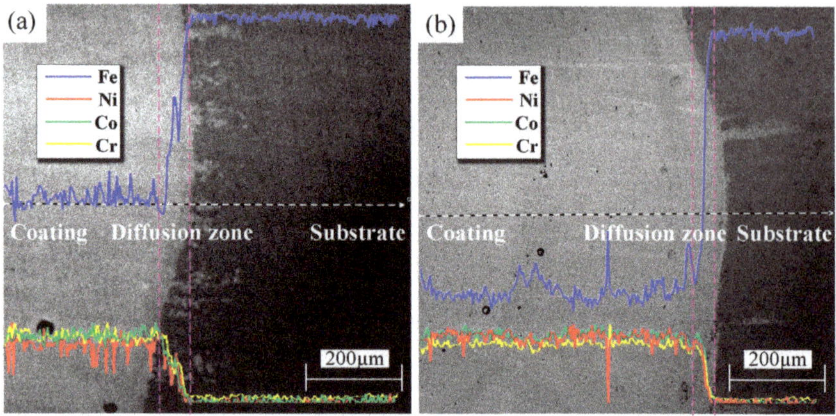

Figure 6. Elements distribution of samples. (**a**) FeCoNiCr, (**b**) CoCrNi.

During the laser cladding process, the difference in the coefficients of thermal expansion between the substrate and laser cladding coating led to thermal stress generation. This caused a greater stress gradient from the substrate material to the cladding layer surface and an uneven distribution of internal stress in the cladding coating. Moreover, the internal pores and defects resulted in the uneven hardness of the cladding coating surface.

Figure 7a–c presents the metallographic structure images of the LC(FeCoNiCr) sample from the top of the coating to the substrate material, respectively. At the juncture of the coating with the substrate, the temperature gradient was the largest, and the grain growth rate was slow, leading to the formation of a plane crystal, as shown in Figure 7c. The cellular dendritic crystal zone was above the plane crystal, and there was a layer of coarse grains of cellular dendritic crystal, with an upward growth perpendicular to the substrate. When the plane crystal formed, the increase in substrate material temperature led to the decrease in the degree of subcooling, and thus the nucleation rate decreased but had little influence on the growth rate of the grains. The preferential growth direction of the grains is consistent with the opposite direction of the fastest heat dissipation direction; that is, upward diffusion from the substrate, which makes the grains grow upwards, perpendicular to the substrate, as shown in Figure 7b. The top grain was cooled in air with a greater degree of undercooling. The growth of columnar crystals was hindered at the solidification front. Moreover, as the degree of supercooling increased to a certain degree, many more new crystal nuclei appeared. They began to grow in all directions, formed an equiaxed-grain structure, and then prevented the growth of a columnar crystal. In addition, the heat flux at the top was greatly affected by the movement of the laser beam, leading to the growth direction of the equiaxed-grain structure parallel to the processing direction, as shown in Figure 7a.

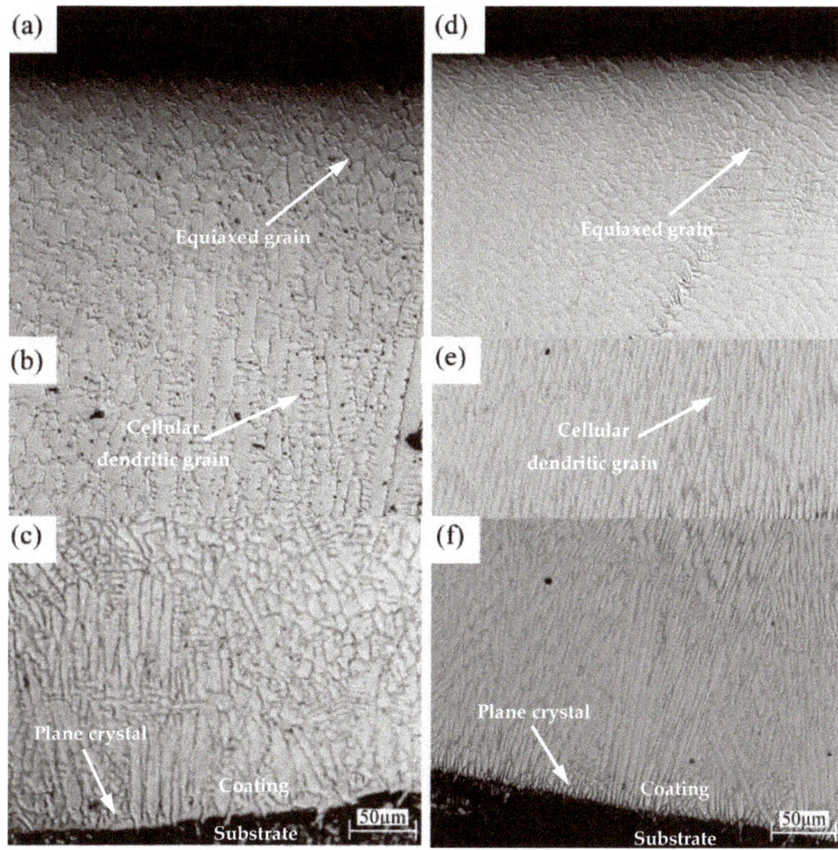

Figure 7. The metallographic images of the cross-sections of coating samples: (**a–c**) for LC(FeCoNiCr); (**d–f**) for LC(CoCrNi).

Figure 7d–f shows the metallographic structure images of the LC(CoCrNi) sample from the top of the coating to the substrate material, respectively. Obviously, the top and bottom of the grain structures of CoCrNi MEA were similar to those of FeCoNiCr MEA. However, during the formation of the plane crystal at the bottom, the temperature gradient in the liquid phase at the interface front of CoCrNi MEA was smaller than that of FeCoNiCr MEA, resulting in a columnar crystal in the middle area rather than a cellular dendritic crystal.

Figure 8 displays the metallographic structures of the longitudinal section of the UB-treated coating samples. It can be seen that the equiaxed-grain structure at the top of the coating disappears, and the fine-grained microstructure forming in the UB-treated coating samples takes its place. The surface influence layer of FeCoNiCr and CoCrNi MEAs is divided into two zones; namely, the grain refinement zone of Zone I and the unaffected zone of Zone II. Zone I refers to the outermost surface layer, in which the FeCoNiCr and CoCrNi MEAs form severe plastic deformation (SPD) layers with a thickness of 41.25 μm and 92.5 μm, respectively, and the grains are obviously refined. In Zone II, there is no obvious change in grain size owing to the greater distance from the processing surface.

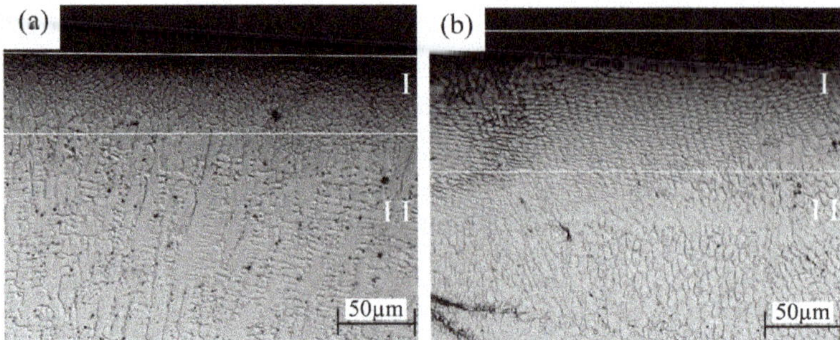

Figure 8. The metallographic images of the cross-sections of coating samples, (**a**) UB(FeCoNiCr), (**b**) UB(CoCrNi).

In the process of UB, the rolling head impacts the coating surface under the co-action of ultrasonic vibration and static pressure, leading to the severe plastic deformation of the coating surface, accompanied by plastic deformations such as crystal plane slip, twinning, and grain boundary migration, the formation of high-density dislocation tangles and dislocation walls, as well as the emergence of grain in a streamlined structure [28]. As a result, the processing quality of the alloy surface is improved with residual compressive stress, and the refined gradient structures are formed in the surface layer. Furthermore, the FeCoNiCr and CoCrNi MEAs have a single FCC structure with more slip systems than BCC and HCP, which is easier to slide to form a streamlined structure in the deformation process [35,36].

3.4. Micro-Hardness and Elastic Modulus

The surface micro-hardness values of the substrate sample and four kinds of coating samples are displayed in Figure 9a. From the test results, the surface micro-hardness values of the LC(FeCoNiCr) and LC(CoCrNi) coatings were 303.5 HV and 340.6 HV, increasing by 45.2% and 62.9%, respectively, in comparison with the substrate. Furthermore, the micro-hardness values of the UB(FeCoNiCr) and UB(CoCrNi) MEA coatings reached 430HV and 452HV, increasing by 41.7% and 32.7%, respectively, compared to their corresponding untreated coatings. Hence, within the experimental scope, UB treatment could significantly improve coating hardness.

Figure 9b presents the cross-section micro-hardness variations of four coatings. It is found that the hardness distributions of the two untreated coatings were relatively even from the top surface to the coating–substrate interface, implying that the microstructures of the two untreated coatings were even. In comparison, the micro-hardness distributions of two UB-treated coatings presented a gradient change from the top surface down to the coating–substrate interface. Meanwhile, in all cases, the highest micro-hardness was distributed at the top surfaces. UB treatment is a kind of severe plastic deformation (SPD) technique. During UB, the top surface materials contact the burnishing roller and are subjected to the most severe deformation, thus having the finest grains, as evident in Figure 8. It is easy to surmise that the material plastic deformation resulting from dislocation by UB was decreasingly severe from the top surface down to the coating–substrate interface. Therefore, as seen in Figure 8, for the two kinds of UB-treated coatings, fine equiaxed grains formed on the top surfaces, and the grain sizes increased in an in-depth direction beneath the top surfaces. According to the Hall–Petch relationship [37,38], finer grains mean higher hardness. Therefore, a gradient hardness distribution formed in the cross-sections of two UB-treated coatings.

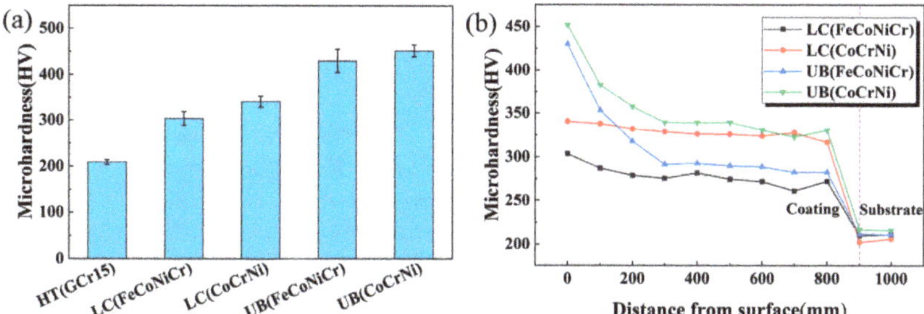

Figure 9. Micro-hardness results. (**a**) Surface hardness; (**b**) cross-section hardness variations.

In order to accurately characterize the mechanical properties of the four studied coatings, a nano-indentation test was carried out. As a result, the loading and unloading load-displacement curves are shown in Figure 10a. During the loading process, there were no obvious differences in the load–displacement curves of the five test coatings with a depth of penetration less than 50 nm, ascribing to the indentation size effect [39]. As the test began, the indenter made contact with partial surface asperities. With the indentation depth increasing, the asperities deformed plastically under the load. After the asperities were completely flattened by the indenter, the nano-indentation behavior of the coating was ultimately determined by its mechanical properties [40]. As shown in Figure 10a, under the maximum load of 30 mN, the maximum indentation depths of the HT(GCr15), LC(FeCoNiCr), LC(CoCrNi), UB(FeCoNiCr), and UB(CoCrNi) coatings were 701 nm, 503 nm, 490 nm, 459 nm, and 444 nm, respectively. Compared with the substrate material (HT(GCr15)), the plastic deformations of four coatings decreased. Meanwhile, the plastic deformations of the two UB-treated coatings were further reduced in comparison with the corresponding untreated coatings. Among them, the indentation depth of the UB(CoCrNi) coating was the minimum. As stated above, by UB treatment, the coating grains were refined, resulting from severe plastic deformation and thus high-density dislocations of materials. As known, finer grains meant more grain boundaries and therefore stronger resistance to deformation. Meanwhile, in the case of small grain size, under the external force, there would occur plastic deformation mostly inside grains, leading to uniform deformation and a small indentation depth.

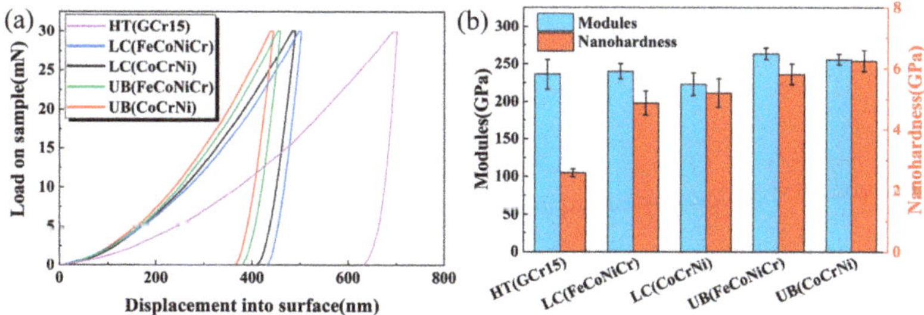

Figure 10. Nano-indentation test results. (**a**) Load–displacement curves; (**b**) elastic moduli and nano-hardness values.

The nano-hardness and elastic modulus obtained through nano-indentation are two important parameters to characterize coating properties. Generally, the bearing capacity and wear resistance of the coating can be reflected by the ratio of the hardness to elastic modulus, i.e., H/E and H^3/E^2, respectively, because they are stronger with the values of H/E and H^3/E^2 increasing [41]. Figure 10b shows the nano-hardness values of different samples. The surface nano-hardnesses of the HT(GCr15), LC(FeCoNiCr), LC(CoCrNi),

UB(FeCoNiCr), and UB(CoCrNi) samples were 2.578 GPa, 4.871 GPa, 5.201 GPa, 5.808 GPa, and 6.259 GPa, respectively. For the two kinds of coating materials, the nano hardness values of UB-treated coatings improved by about 20% in comparison with their corresponding untreated coatings, which was in good agreement with the results of the micro-Vickers hardness test in Figure 9a. The elastic moduli of different samples are presented in Figure 10b. Similar observations can be found for the two kinds of coating materials. That is, although there was a slight increase in the elastic modulus after UB treatment, such an increase was much less significant in comparison with that in nano-hardness. As can be seen from the test results, the grain refinement and residual stress induction caused by UB treatment played a significant role in the increase in coating hardness. However, the coating elastic modulus was not obviously changed.

3.5. Friction Wear

Figure 11a,b displays friction coefficient variations against time, as well as the average friction coefficients of the five different samples, respectively. As exemplified in Figure 11a, the friction coefficients of the five kinds of samples experienced a short running-in period at the very beginning of the wear test with rapid rising, then reached a steady state. It can be seen that when compared with the substrate sample, the friction coefficient fluctuations of four coatings decreased to varying degrees. In particular, the UB(CoCrNi) coating had the shortest run-in period and the minimum fluctuation. As shown in Figure 11b, the average friction coefficients of the HT(GCr15), LC(FeCoNiCr), LC(CoCrNi), UB(FeCoNiCr), and UB(CoCrNi) samples were 0.592, 0.503, 0.456, 0.410, and 0.375, respectively. Among them, the friction coefficient of the UB(CoCrNi) sample was the lowest. As stated above, there was grain refinement and surface work hardening after the UB treatment. Due to the surface smoothening and hardening effect resulting from UB, during wear, the contact area between the counterpart and coating surface became small, thus decreasing the friction coefficient [42,43].

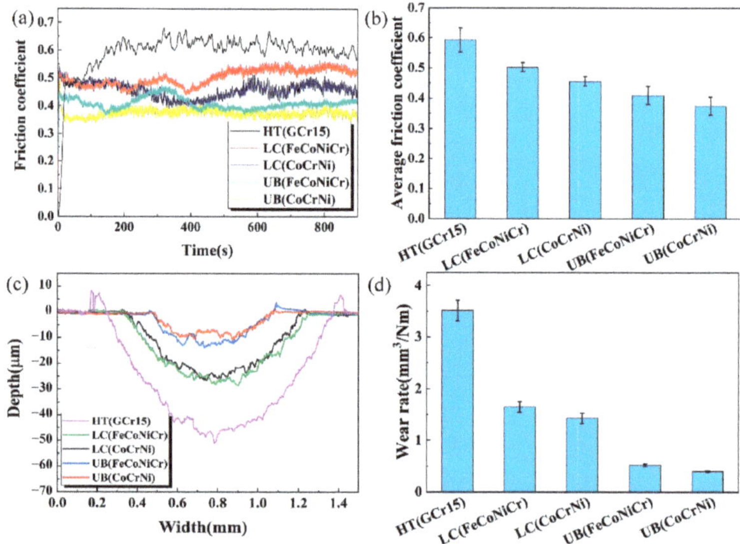

Figure 11. Friction test results. (**a**) Friction coefficient variations; (**b**) average friction coefficient; (**c**) wear profile curve of the cross-section; (**d**) wear rates of samples.

Figure 11c,d shows the cross-section profiles of the wear scars and wear rates of the five tested samples. In Figure 11c, there formed bulges on both sides of the wear traces of the substrate sample and the two untreated coating samples, suggesting that extrusion and plastic flow occurred during wear. It is clear that the wear trace of the substrate sample had the highest bulges. Notably, the two UB-treated coating samples almost had no bulges, ascribing to the work-hardening effect of the UB treatment. As such, the metal flow was weaker after work hardening, with a difficulty in plastic deformation [8]. In terms of wear size, the substrate sample had the most serious wear, as well as the largest wear width and depth among all the samples. The wear width and depth of two untreated coating samples became smaller. Similarly, the wear widths and depths of the two UB-treated coating samples were further reduced in comparison to those without UB treatment, implying that the UB-treated samples had better wear resistance. The wear rates of all samples were calculated and are presented in Figure 11d. As can be seen, the wear rates of FeCoNiCr and CoCrNi untreated coatings were 1.65 mm^3/Nm and 1.43 mm^3/Nm, respectively, so their wear resistances were close to each other. Compared with the wear rate of the substrate sample (3.51 mm^3/Nm), they decreased by 53.0% and 59.3%, respectively. Likewise, the wear rates of the UB-treated FeCoNiCr and CoCrNi coatings were 0.52 mm^3/Nm and 0.40mm^3/Nm, respectively, and decreased by 68.5% and 72.0% in comparison to those before UB treatment, illustrating that the UB treatment had a role in reducing the friction coefficients of FeCoNiCr and CoCrNi coatings. According to Archard's law, wear rate is inversely proportional to hardness [44], so high hardness, as well as a large elastic modulus, were beneficial to improving the wear resistance of the FeCoNiCr and CoCrNi coatings.

Figure 12a–e presents the SEM images of the wear areas of five kinds of samples. As shown in Figure 12a, the substrate sample showed obvious wear trace, with the main characteristics of micro-cracks, deep grooves, delamination, and wear debris. In the process of friction, micro-cracks easily occurred as the shear stress induced by friction exceeded the yield strength of the sample. The delamination mainly resulted from micro-crack propagation and coalescence. Furthermore, during the wear process, under the co-action of the normal load and shear force of the friction pair, the wear debris exfoliated from the sample surface, causing adhesive wear. The exfoliated wear debris embedded in the sample surface under the normal load. Later, the surface was scratched under shear load, thus accelerating the abrasive wear and then forming the parallel deep grooves. Hence, it is suggested from the wear characteristics that the wear mechanism of the substrate sample was mainly abrasive wear and adhesive wear.

As can be seen from Figure 12b,c, compared with the substrate sample, there was no visible crack found in the wear area of the LC(FeCoNiCr) and LC(CoCrNi) samples, and the delamination and grooves were obviously alleviated. Moreover, the size of the wear debris decreased. The reason for the wear resistance improvement of the untreated coating samples was the high surface hardness. A hard surface could hinder the deformation of the material, reducing micro-cracks as well as weakening the delamination wear. It can be found from Figure 12d,e that the wear of the two UB-treated coating samples was greatly weakened and the quantity and volume of wear debris in the wear areas decreased, and the grooves were shallowed compared with their corresponding untreated coating samples. The grains of the two UB-treated coating samples were refined via UB treatment, and their hardness and yield strengths were significantly improved. Thereby, the volume and quantity of wear debris produced in the wear test were reduced, and the abrasive wear was alleviated [45,46].

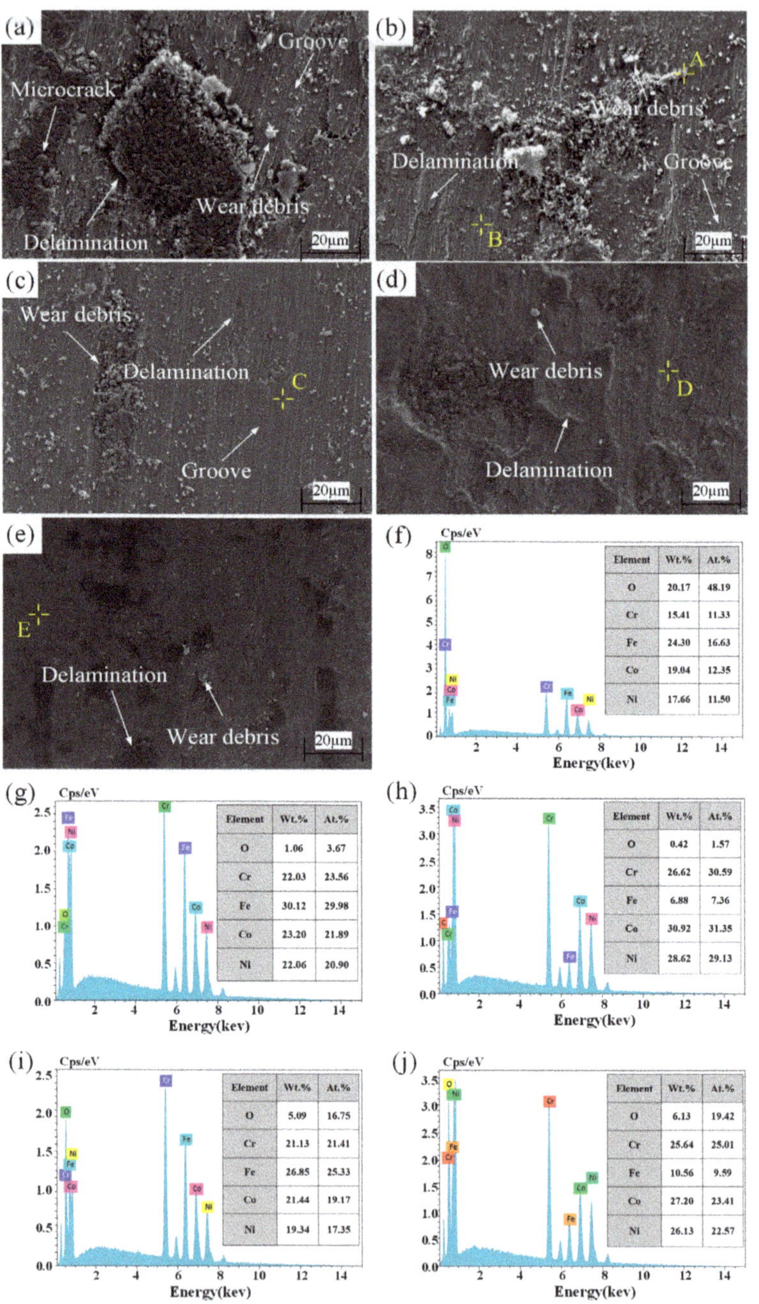

Figure 12. SEM images of wear areas and EDS analysis. (**a**) HT(GCr15), (**b**) LC(FeCoNiCr), (**c**) LC(CoCrNi), (**d**) UB(FeCoNiCr), (**e**) UB(CoCrNi), (**f**) element analysis of wear debris A of LC(FeCoNiCr), (**g**) element analysis of wear scar B of LC(FeCoNiCr), (**h**) element analysis of wear scar C of LC(CoCrNi), (**i**) element analysis of wear scar D of UB(FeCoNiCr), and (**j**) element analysis of wear scar E of UB(CoCrNi).

In order to determine whether there was oxidation wear in the wear process, the EDS analysis was carried out for the wear debris A and wear scars B, C, D, and E of the coating samples, as shown in Figure 12b–e. The element contents at A–E are illustrated in Figure 12f–j, and clearly, there was oxidation wear at all points A–E. During the oxidation wear, the oxidation film generated was ductile and softer than the substrate. Under the shear force between the friction pair, partial oxidation films would be exfoliated, and the exfoliated films could be smeared on the sample surface, serving as a solid lubricant and thereby reducing wear [47]. As shown in Figure 12f,g, the O element content of wear debris A was obviously higher than that at B without wear debris, which confirmed that the oxidation film was continuously exfoliated to form wear debris. In Figure 12g–j, the O element content at D and E was much higher than that at B and C, respectively, suggesting that the oxidation wear of the UB-treated coatings was more obvious. This is because the grain refinement, dislocation, and high internal stress generated from the UB treatment could make the oxidation film denser and thicker [48]. In addition, according to Archard's law, the larger the load was in the wear process, the higher the hardness and the higher the temperature of the sample surface [49]. More importantly, oxidation wear easily occurs under high temperatures [50]. Hence, the UB treatment could enhance the oxidation wear, beneficial to improving the wear resistance of the coating surface.

4. Conclusions

This study adopted UB treatment for the surface strengthening of FeCoNiCr and CoCrNi coatings prepared by laser cladding. The results paved the way for the preparation and post-treatment of FeCoNiCr and CoCrNi coatings. By analysis, the conclusions were as follows:

(1) In comparison with the LC(FeCoNiCr) and LC(CoCrNi) coating samples, the surface roughness and porosity of the two corresponding UB-treated samples decreased significantly, illustrating that UB treatment could greatly smoothen the coating surface, decrease the porosity of coatings, and reduce surface defects. During UB treatment, under the applied external dynamic load, the materials in the coating surface and near-surface were forced to flow from surface peaks to valleys, and thereby the coating surface was flattened and the pores and defects in the coating surface were filled/cured.

(2) The Fe content in two kinds of coatings obviously increased due to the dilution effect of substrate on the MEAs during laser cladding. The microstructure of the two kinds of MEA coatings showed a plane crystal at the bottom, a columnar crystal (cellular dendritic crystal for CoCrNi coating) at the middle, and an equiaxed-grain crystal at the top. For the two kinds of coating, after UB treatment, there was obvious grain refinement within 100 μm beneath the top surfaces of the coatings, which resulted from dislocation accumulation from the severe plastic deformation of the materials.

(3) Compared with the LC(FeCoNiCr) and LC(CoCrNi) samples, the surface hardness and yield strength of the two corresponding UB-treated coating samples were improved. In both cases, after UB treatment, a gradient hardness structure was formed along the in-depth direction of the coating. Meanwhile, the two kinds of UB-treated coating samples exhibited better friction and wear properties than their corresponding untreated samples with wear rates decreasing by 68.5% and 72.0%, respectively.

Author Contributions: Conceptualization: X.S. and B.W. Experiment design, material preparation, and data collection were performed by C.L. and B.W. The first draft of the manuscript was written by C.L. Writing—review and editing by X.S. and B.W. Formal analysis, G.S. and Y.Z. Investigation, G.S. and A.L.; visualization, A.L.; project administration, Y.Z. All authors have read and agreed to the published version of the manuscript.

Funding: This study was funded by the National Natural Science Foundation of China (grant numbers 51775285 and 52075275).

Institutional Review Board Statement: Not applicable.

Informed Consent Statement: Not applicable.

Data Availability Statement: The data presented in this study are available from the corresponding author upon reasonable request.

Conflicts of Interest: The authors declare no conflict of interest.

References

1. Yeh, J.-W.; Chen, S.K.; Lin, S.-J.; Gan, J.-Y.; Chin, T.-S.; Shun, T.-T.; Tsau, C.-H.; Chang, S.-Y. Nanostructured High-Entropy Alloys with Multiple Principal Elements: Novel Alloy Design Concepts and Outcomes. *Adv. Eng. Mater.* **2004**, *6*, 299–303. [CrossRef]
2. Motallebi, R.; Savaedi, Z.; Mirzadeh, H. Superplasticity of high-entropy alloys: A review. *Arch. Civ. Mech. Eng.* **2021**, *22*, 20. [CrossRef]
3. Pandian, V.; Kannan, S. Effect of high entropy particle on aerospace-grade aluminium composite developed through combined mechanical supersonic vibration and squeeze infiltration technique. *J. Manuf. Process.* **2022**, *74*, 383–399. [CrossRef]
4. Pang, J.; Zhang, H.; Zhang, L.; Zhu, Z.; Fu, H.; Li, H.; Wang, A.; Li, Z.; Zhang, H. A ductile Nb40Ti25Al15V10Ta5Hf3W2 refractory high entropy alloy with high specific strength for high-temperature applications. *Mater. Sci. Eng. A* **2022**, *831*, 142290. [CrossRef]
5. Guo, Y.; Liu, L.; Zhang, W.; Yao, K.; Chen, W.; Ren, J.; Qi, J.; Wang, B.; Zhao, Z.; Shang, J.; et al. A new method for preparing high entropy alloys: Electromagnetic pulse treatment and its effects on mechanical and corrosion properties. *Mater. Sci. Eng. A* **2020**, *774*, 138916. [CrossRef]
6. Yeh, J.-W. Recent progress in high-entropy alloys. *Eur. J. Control* **2006**, *31*, 633–648. [CrossRef]
7. Cantor, B.; Chang, I.T.H.; Knight, P.; Vincent, A.J.B. Microstructural development in equiatomic multicomponent alloys. *Mater. Sci. Eng. A* **2004**, *375–377*, 213–218. [CrossRef]
8. Tong, Z.; Pan, X.; Zhou, W.; Yang, Y.; Ye, Y.; Qian, D.; Ren, X. Achieving excellent wear and corrosion properties in laser additive manufactured CrMnFeCoNi high-entropy alloy by laser shock peening. *Surf. Coat. Technol.* **2021**, *422*, 127504. [CrossRef]
9. Ye, H.; Zhu, J.; Liu, Y.; Liu, W.; Wang, D. Microstructure and mechanical properties of laser cladded Cr Ni alloy by hard turning (HT) and ultrasonic surface rolling (USR). *Surf. Coat. Technol.* **2020**, *393*, 125806. [CrossRef]
10. Long, Y.; Yang, J.; Peng, H.; Vogel, F.; Chen, W.; Su, K. Combination of enhanced strength and sufficient tensile ductility in a sintered ultrafine-grained CoFeMnNi medium-entropy alloy. *Mater. Sci. Eng. A* **2022**, *831*, 142175. [CrossRef]
11. Gludovatz, B.; Hohenwarter, A.; Thurston, K.V.S.; Bei, H.; Wu, Z.; George, E.P.; Ritchie, R.O. Exceptional damage-tolerance of a medium-entropy alloy CrCoNi at cryogenic temperatures. *Nat. Commun.* **2016**, *7*, 10602. [CrossRef] [PubMed]
12. Sun, D.; Cai, Y.; Zhu, L.; Gao, F.; Shan, M.; Manladan, S.M.; Geng, K.; Han, J.; Jiang, Z. High-temperature oxidation and wear properties of TiC-reinforced CrMnFeCoNi high entropy alloy composite coatings produced by laser cladding. *Surf. Coatings Technol.* **2022**, *438*, 128407. [CrossRef]
13. Wang, J.; Zhang, C.; Shen, X.; He, J. A study on surface integrity of laser cladding coatings post-treated by ultrasonic burnishing coupled with heat treatment. *Mater. Lett.* **2022**, *308*, 131136. [CrossRef]
14. Zhang, M.; Zhou, X.; Yu, X.; Li, J. Synthesis and characterization of refractory TiZrNbWMo high-entropy alloy coating by laser cladding. *Surf. Coat. Technol.* **2017**, *311*, 321–329. [CrossRef]
15. Liang, H.; Yao, H.; Qiao, D.; Nie, S.; Lu, Y.; Deng, D.; Cao, Z.; Wang, T. Microstructures and Wear Resistance of AlCrFeNi$_2$V$_{0.5}$Tix High-Entropy Alloy Coatings Prepared by Laser Cladding. *J. Therm. Spray Technol.* **2019**, *28*, 1318–1319. [CrossRef]
16. Lu, K. Making strong nanomaterials ductile with gradients. *Science* **2014**, *345*, 1455–1456. [CrossRef]
17. Shen, X.; Su, H.; Wang, J.; Zhang, C.; Xu, C.; Bai, X. New approach towards the machining process after laser cladding. *Arch. Civ. Mech. Eng.* **2021**, *21*, 8. [CrossRef]
18. Shi, Y.-L.; Shen, X.-H.; Xu, G.-F.; Xu, C.-H.; Wang, B.-L.; Su, G.-S. Surface integrity enhancement of austenitic stainless steel treated by ultrasonic burnishing with two burnishing tips. *Arch. Civ. Mech. Eng.* **2020**, *20*, 79. [CrossRef]
19. Liu, Z.; Yang, M.; Deng, J.; Zhang, M.; Dai, Q. A predictive approach to investigating effects of ultrasonic-assisted burnishing process on surface performances of shaft targets. *Int. J. Adv. Manuf. Technol.* **2020**, *106*, 4203–4219. [CrossRef]
20. Meng, Y.; Deng, J.; Wu, J.; Wang, R.; Lu, Y. Improved interfacial adhesion of AlTiN coating by micro-grooves using ultrasonic surface rolling processing. *J. Mater. Process. Technol.* **2022**, *304*, 117570. [CrossRef]
21. Luo, X.; Ren, X.; Jin, Q.; Qu, H.; Hou, H. Microstructural evolution and surface integrity of ultrasonic surface rolling in Ti6Al4V alloy. *J. Mater. Res. Technol.* **2021**, *13*, 1586–1598. [CrossRef]
22. Dong, P.; Liu, Z.; Zhai, X.; Yan, Z.; Wang, W.; Liaw, P.K. Incredible improvement in fatigue resistance of friction stir welded 7075-T651 aluminum alloy via surface mechanical rolling treatment. *Int. J. Fatigue* **2019**, *124*, 15–25. [CrossRef]
23. Xia, T.; Zeng, L.; Zhang, X.; Liu, J.; Zhang, W.; Liang, T.; Yang, B. Enhanced corrosion resistance of a Cu 10Ni alloy in a 3.5 wt.% NaCl solution by means of ultrasonic surface rolling treatment. *Surf. Coat. Technol.* **2019**, *363*, 390–399. [CrossRef]
24. Giusca, C.L.; Leach, R.K.; Helary, F.; Gutauskas, T.; Nimishakavi, L. Calibration of the scales of areal surface topography-measuring instruments: Part 1. Measurement noise and residual flatness. *Meas. Sci. Technol.* **2012**, *23*, 035008. [CrossRef]
25. Podulka, P. Suppression of the High-Frequency Errors in Surface Topography Measurements Based on Comparison of Various Spline Filtering Methods. *Materials* **2021**, *14*, 5096. [CrossRef]

26. Haitjema, H. Uncertainty in measurement of surface topography. *Surf. Topogr. Metrol. Prop.* **2015**, *3*, 035004. [CrossRef]
27. Podulka, P. The Effect of Surface Topography Feature Size Density and Distribution on the Results of a Data Processing and Parameters Calculation with a Comparison of Regular Methods. *Materials* **2021**, *14*, 4077. [CrossRef]
28. Boryczko, A. Effect of waviness and roughness components on transverse profiles of turned surfaces. *Measurement* **2013**, *46*, 688–696. [CrossRef]
29. Raja, J.; Muralikrishnan, B.; Fu, S. Recent advances in separation of roughness, waviness and form. *Precis. Eng.* **2002**, *26*, 222–235. [CrossRef]
30. Ye, H.; Sun, X.; Liu, Y.; Rao, X.-X.; Gu, Q. Effect of ultrasonic surface rolling process on mechanical properties and corrosion resistance of AZ31B Mg alloy. *Surf. Coat. Technol.* **2019**, *372*, 288–298. [CrossRef]
31. Shen, X.; Gong, X.; Wang, B.; He, J.; Xu, C.; Su, G. Surface properties enhancement of Inconel 718 alloy by ultrasonic roller burnishing coupled with heat treatment. *Arch. Civ. Mech. Eng.* **2021**, *21*, 122. [CrossRef]
32. Liu, K.; Gu, D.; Guo, M.; Sun, J. Effects of processing parameters on densification behavior, microstructure evolution and mechanical properties of W–Ti alloy fabricated by laser powder bed fusion. *Mater. Sci. Eng. A* **2022**, *829*, 142177. [CrossRef]
33. Zhang, C.; Shen, X.; Wang, J.; Xu, C.; He, J.; Bai, X. Improving surface properties of Fe-based laser cladding coating deposited on a carbon steel by heat assisted ultrasonic burnishing. *J. Mater. Res. Technol.* **2021**, *12*, 100–116. [CrossRef]
34. Yang, J.; Bai, B.; Ke, H.; Cui, Z.; Liu, Z.; Zhou, Z.; Xu, H.; Xiao, J.; Liu, Q.; Li, H. Effect of metallurgical behavior on microstructure and properties of FeCrMoMn coatings prepared by high-speed laser cladding. *Opt. Laser Technol.* **2021**, *144*, 107431. [CrossRef]
35. Zhang, F.; Wu, Y.; Lou, H.; Zeng, Z.; Prakapenka, V.B.; Greenberg, E.; Ren, Y.; Yan, J.; Okasinski, J.S.; Liu, X.; et al. Polymorphism in a high-entropy alloy. *Nat. Commun.* **2017**, *8*, 15687. [CrossRef]
36. Otto, F.; Dlouhý, A.; Somsen, C.; Bei, H.; Eggeler, G.; George, E.P. The influences of temperature and microstructure on the tensile properties of a CoCrFeMnNi high-entropy alloy. *Acta Mater.* **2013**, *61*, 5743–5755. [CrossRef]
37. Wang, H.; Song, G.; Tang, G. Effect of electropulsing on surface mechanical properties and microstructure of AISI 304 stainless steel during ultrasonic surface rolling process. *Mater. Sci. Eng. A* **2016**, *662*, 456–467. [CrossRef]
38. Wang, K.Y.; Shen, T.D.; Quan, M.X.; Wei, W.D. Hall-Petch relationship in nanocrystalline titanium produced by ball-milling. *J. Mater. Sci. Lett.* **1993**, *12*, 1818–1820. [CrossRef]
39. Qin, L.; Li, H.; Shi, X.; Beake, B.D.; Xiao, L.; Smith, J.F.; Sun, Z.; Chen, J. Investigation on dynamic hardness and high strain rate indentation size effects in aluminium (110) using nano-impact. *Mech. Mater.* **2019**, *133*, 55–62. [CrossRef]
40. Dao, M.; Lu, L.; Asaro, R.; De Hosson, J.; Ma, E. Toward a quantitative understanding of mechanical behavior of nanocrystalline metals. *Acta Mater.* **2007**, *55*, 4041–4065. [CrossRef]
41. Liu, Y.; Wang, Q.; Xie, J.; Yang, X.; Peng, P.; Wang, Y.; Li, M.; Ryu, C.H.; Joo, Y.H.; Jeong, I.W.; et al. Enhanced surface composite coating on Ti811 alloy by laser cladding towards improved nano-hardness. *Ceram. Int.* **2022**, *48*, 18773–18783. [CrossRef]
42. Wang, H.-T.; Yao, H.-L.; Zhang, M.-X.; Bai, X.-B.; Yi, Z.-H.; Chen, Q.-Y.; Ji, G.-C. Surface nanocrystallization treatment of AZ91D magnesium alloy by cold spraying shot peening process. *Surf. Coat. Technol.* **2019**, *374*, 485–492. [CrossRef]
43. Hou, G.; Li, A. Effect of Surface Micro-Hardness Change in Multistep Machining on Friction and Wear Characteristics of Titanium Alloy. *Appl. Sci.* **2021**, *11*, 7471. [CrossRef]
44. Lakshminarayanan, R.; Chao, L.-Y.; Iyer, N.; Shetty, D. Wear of steel in rolling contact with silicon nitride. *Wear* **1997**, *210*, 278–286. [CrossRef]
45. Yingxia, Y.; Bolin, H.; Zongmin, L.; Siyong, L.; Songsong, X. Experimental Research on Microhardness and Wear Resistance of MB8 Magnesium Alloy Treated by Ultrasonic Impact. *Rare Met. Mater. Eng.* **2017**, *46*, 1798–1802. [CrossRef]
46. Su, Y.; Wang, Z.; Lu, H.; Luo, K.; Lu, J. Improved wear resistance of directed energy deposited Fe-Ni-Cr alloy via closed-loop controlling laser power. *J. Manuf. Process.* **2022**, *75*, 802–813. [CrossRef]
47. Avcu, E. The influences of ECAP on the dry sliding wear behaviour of AA7075 aluminium alloy. *Tribol. Int.* **2017**, *110*, 173–184. [CrossRef]
48. Jiang, J.; Ma, A.; Song, D.; Yang, D.; Shi, J.; Wang, K.; Zhang, L.; Chen, J. Anticorrosion behavior of ultrafine-grained Al-26 wt% Si alloy fabricated by ECAP. *J. Mater. Sci.* **2012**, *47*, 7744–7750. [CrossRef]
49. Archard, J. The temperature of rubbing surfaces. *Wear* **1959**, *2*, 438–455. [CrossRef]
50. Xia, S.; Liu, Y.; Fu, D.; Jin, B.; Lu, J. Effect of Surface Mechanical Attrition Treatment on Tribological Behavior of the AZ31 Alloy. *J. Mater. Sci. Technol.* **2016**, *32*, 1245–1252. [CrossRef]

Article

Effect of Cr on Microstructure and Properties of WVTaTiCr$_x$ Refractory High-Entropy Alloy Laser Cladding

Zhaomin Xu, Zhiping Sun *, Cheng Li and Zhiming Wang

School of Mechanical Engineering, Qilu University of Technology (Shandong Academy of Sciences), Jinan 250353, China
* Correspondence: benzsun@163.com

Abstract: WVTaTiCr$_x$ (x = 0, 0.25, 0.5, 0.75, 1) refractory high-entropy alloy coatings were prepared on a 42-CrMo steel plate using laser cladding. The purpose of this work is to investigate the effect of the Cr content on the microstructure and properties of the WVTaTiCr$_x$ coating. The morphologies and phase compositions of five coatings with different Cr contents were comparatively observed. In addition, the hardness and high-temperature oxidation resistance of the coatings were also analyzed. As a result, with the increase in Cr, the coating grains were more refined. All the coating is mainly composed of the BCC solid-solution phase, which promotes the precipitation of the Laves phase with the increase in Cr. The addition of Cr greatly improves the hardness, high-temperature oxidation resistance and corrosion resistance of the coating. The WVTaTiCr (Cr$_1$) exhibited superior mechanical properties, especially in terms of its exceptional hardness, high-temperature oxidation resistance and outstanding corrosion resistance. The average hardness of the WVTaTiCr alloy coating reaches 627.36 HV. After 50 h of high-temperature oxidation, the oxide weight of WVTaTiCr increases by 5.12 mg/cm^2, and the oxidation rate is 0.1 mg/(cm^2·h). In 3.5 wt% NaCl solution, the corrosion potential of WVTaTiCr is −0.3198 V, and the corrosion rate is 0.161 mm/a.

Keywords: refractory high-entropy alloys; laser cladding; microstructure; mechanical properties; oxidation resistance

Citation: Xu, Z.; Sun, Z.; Li, C.; Wang, Z. Effect of Cr on Microstructure and Properties of WVTaTiCr$_x$ Refractory High-Entropy Alloy Laser Cladding. *Materials* **2023**, *16*, 3060. https://doi.org/10.3390/ma16083060

Academic Editor: Norbert Robert Radek

Received: 30 March 2023
Revised: 8 April 2023
Accepted: 11 April 2023
Published: 13 April 2023

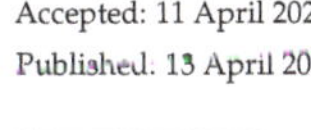

1. Introduction

Refractory high-entropy alloys (RHEAs) first appeared in public view in 2010 [1]. Thereafter, they aroused great concern from researchers at home and abroad. Research has proved that RHEAs have excellent mechanical properties at high temperatures, even surpassing the widely used nickel-based alloy [2–6]. RHEAs, with excellent mechanical properties at elevated temperatures, are considered promising high-temperature application materials, which are suitable for complex working conditions such as atomic energy, aerospace, the military industry, advanced nuclear reactors, etc. Previous studies focused on mechanical properties at high- and room-temperatures, but with little research and analysis on high-temperature oxidation resistance [7–9]. The research shows that adding the proper amount of Al, Cr, and Si elements into the refractory high-entropy alloy can form a dense oxidation film on the surface of the alloy in a high-temperature environment, thus improving the high-temperature oxidation resistance [10–13]. Unfortunately, elements Al, Cr, and Si very easily form brittle intermetallic compounds with other refractory elements, resulting in reduced mechanical properties of RHEAs [14–16]. As a "candidate" for new high-temperature alloy materials, refractory high-entropy alloys still have top priority in improving high-temperature oxidation resistance.

In recent years, W-based alloys have been widely used in the nuclear energy industry, and W-Ta [17], W-V [18], W-Ti [19], W-Cr [20] have been studied. Research results showed that some W-based binary alloys showed brittle behavior [21]. The reduction in ductility makes it difficult for W-based binary alloys to be widely used in harsh working environments, which also urges us to shift our research focus to W- based high-entropy alloys, providing a new idea for the development of heat insulation and radiation protection materials for nuclear reactors. W and Ta elements have adequate mechanical properties and resistance to irradiation-induced embrittlement and swelling [22]. Ti plays a significant role in improving the sintered density [19], whereas V aids in improving the strength and hardness of refractory HEAs [23]. Another refractory metal, Cr was also chosen for use, considering its ability for high-temperature oxidation [24–27]. The high-entropy alloys undergo segregation when prepared by melting, and embrittlement and porosity are attributes of conventional sintering, whereas the low thickness of the final product is a limit of the physical vapor deposition (PVD) approach [28]. In addition, electrodeposition techniques lead to pores and microcracks in the coating [29]. With the progress of science and technology and the continuous optimization of process technology, laser cladding technology has gradually appeared in public view and has become a hot research topic [30]. Laser cladding, an advanced surface modification technology, cures metal powder into the substrate or chosen material surface, utilizing a high-power laser [31–33]. Compared with magnetron sputtering, PTA welding, and additive manufacturing, the research shows that the cladding layer forms firm metallurgical bonds with the substrate, the heat-affected zone of the coating is small, the deformation of the substrate is small, and the coating thickness can reach the millimeter level [30]. Based on comprehensive consideration, W, V, Ta, Ti, and Cr elements are used to develop a new refractory high-entropy alloy with both hardness and high-temperature oxidation resistance. In this study, $WVTaTiCr_x$ RHEAs coating was prepared on the 42-CrMo steel surface using laser cladding, and the trends in the coating microstructure, hardness, high-temperature oxidation resistance, and corrosion resistance were analyzed by varying the Cr content.

2. Materials and Methods

The 42-CrMo steel was selected as the substrate. The laser cladding powder was W, V, Ta, Ti, and Cr, with the particle size of 45–105μm and powder purity of 99.9%. Table 1 shows the basic properties of the five elements. Five kinds of powder were mixed using ball milling, in a specific proportion. The mass fraction of each element is shown in Table 2, abbreviated as Cr_0, $Cr_{0.25}$, $Cr_{0.5}$, $Cr_{0.75}$, and Cr_1 for the five alloy coatings. The powder was mixed fully in a planetary ball mill. The rotating speed of the ball mill was 300 r/min, and the ball-to-material ratio was 10:1. After 24 h of uninterrupted ball milling, the powder was dried and stored under a vacuum.

Table 1. Basic properties of elements in $WVTaTiCr_x$ RHEA coating [34].

Element	W	V	Ta	Ti	Cr
Atomic radius (Å)	1.37	1.32	1.43	1.46	1.25
Melting point (°C)	3407	1890	2996	1660	1857

Table 2. The mass fraction of each element of alloy coatings (wt%).

Alloys	Identification	W	V	Ta	Ti	Cr
WVTaTi	Cr_0	39.65	10.99	39.03	10.33	
$WVTaTiCr_{0.25}$	$Cr_{0.25}$	38.57	10.69	37.97	10.12	2.74
$WVTaTiCr_{0.5}$	$Cr_{0.5}$	37.55	10.40	36.96	9.78	5.31
$WVTaTiCr_{0.75}$	$Cr_{0.75}$	36.58	10.14	36	9.52	7.76
$WVTaTiCr_1$	Cr_1	35.66	9.88	35.09	9.28	10.08

This cladding experiment was completed on the LYHS series ultra-high-speed laser-cladding machine tool, whose workbench can move the X-Y-Z three axes with multiple degrees of freedom. In this experiment, the presetting method of coating was used, and argon gas was used as a protective gas. After several parameter adjustments, the optimum process parameters were as follows: the laser power was 1.4 kw, the scanning speed was 8 mm/s, the spot diameter was 6 mm, and the overlap rate was 30–40%.

After cladding, the coating surfaces were polished with sandpaper, then corroded with aqua regia (the ratio of concentrated nitric acid to the concentrated hydrochloric acid solution was 1:3), and cleaned with absolute ethanol. A scanning electron microscope (SEM, Phenom ProX, Phenom-World, Eindhoven, The Netherlands) and its accompanying energy dispersive analyzer (EDS) were used to observe and measure the surface morphology, metallographic structure, and chemical composition content of the coatings. The Smart Lab X-ray diffractometer (Rigaku, Tokyo, Japan) (anode Cu target, X-ray Kα, X-ray wavelength 0.1542 nm, and scanning rate ranging from 1°/min to 20°/min) was used for phase analysis of the coating.

The hardness changes in the samples were analyzed with HXD-1000TMC microhardness tester under a load of 1.96 N and a loading time of 15 s. Hardness tests were carried out every 0.1 mm along the vertical direction of the interface between the substrate and the cladding layer, and the hardness changes in the coatings were recorded. The original quality of the sample was weighed with an electronic balance (accuracy 0.0001 g) and recorded. The sample was put into the corundum ceramic boat, the ceramic boat was placed in the center of the muffle furnace, and the heating rate set at 5 °C/min and the target temperature at 800 °C. After taking out the sample every 10 h, it was air-cooled quickly, weighed, and the quality change recorded. The whole high-temperature oxidation experiment lasted 50 h. A dynamic polarization electrochemical test in the 3.5 wt% NaCl aqueous solution was conducted with a potentiostat (Gamary Interface 1000, Warminster, PA, USA) to evaluate the corrosion behaviors of the RHEA coatings. The cladding layer was the working electrode, and the auxiliary and reference electrodes were platinum and saturated calomel, respectively. The corrosion current density (I_{corr}), corrosion potential (E_{corr}), and corrosion rate (V) were determined using the Tafel analysis method.

3. Results and Discussion

3.1. Microstructure Characterization

WVTaTiCr$_x$ (x = 0, 0.25, 0.5, 0.75, 1) RHEA coatings were prepared under the optimum process parameters, and the cladding condition of the coating cross section was observed for WVTaTi(Cr$_0$) as an example, and the results are shown in Figure 1. From Figure 1a, it is observed that the thickness of the coatings is about 1 mm and the thickness of the heat-affected zone (HAZ) is 0.3~0.5 mm. From Figure 1b,c, it is seen that the metallurgical bonding between the 42CrMo steel and the coating is good, and there are no obvious cracks and pores near the bonding zone (BZ). In addition, the typical dendrite structures are formed near the interface between the coating and the substrate, and grow perpendicular to the boundary.

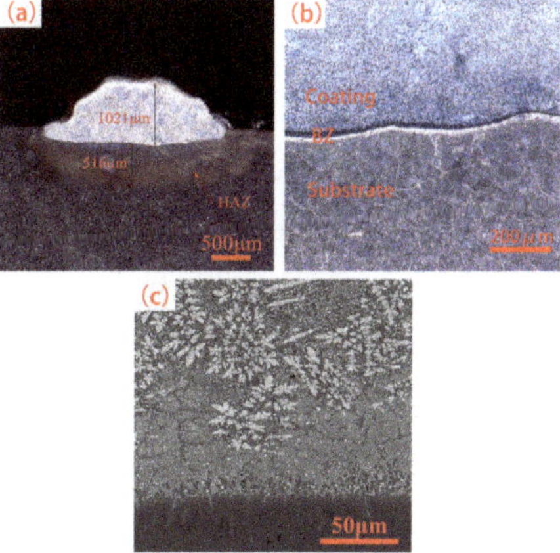

Figure 1. The cladding condition of the coating cross-section. (**a**) coating cross-section; (**b**) the binding zone at low multiples; (**c**) the binding zone at high multiples.

Figure 2 shows the microscopic morphology of different coatings. It can be seen from the diagram that the structural morphology of the cladding layer changes obviously with the change in element content. As shown in Figure 2a,b, when there is no, or only a small amount of, Cr in the coating, the coating mainly consists of white petal-like dendrites and a gray base. As the elemental Cr was added to the coating, the dendritic structure became uniform. From Figure 2d, it can be seen that feather-like structures were generated in the coating. In Figure 2e, it is observed that the structure of Cr_1 refines and the area of feather-like tissue increases, showing a network distribution within the dendritic region.

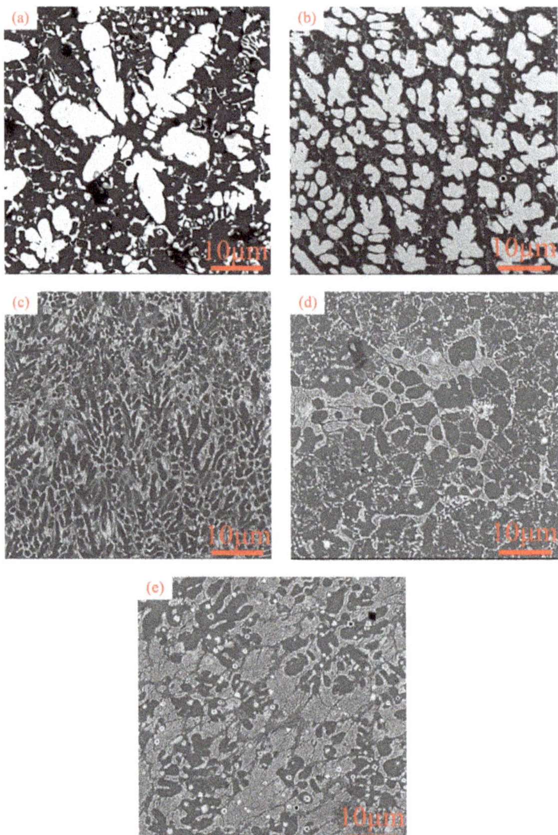

Figure 2. SEM of the cross-section of $WVTaTiCr_x$ refractory high-entropy alloy coating: (**a**) Cr_0; (**b**) $Cr_{0.25}$; (**c**) $Cr_{0.5}$; (**d**) $Cr_{0.75}$; (**e**) Cr_1.

Figure 3 shows the EDS element distribution of the $WVTaTiCr_x$ RHEA coating. The figure shows that during the solidification process of the alloy powder, the distribution of elements in the cladding layer is not uniform. Ta and W with high melting points are first solidified and formed, and mainly distributed in the white dendrite area; there is a high content of V and Cr elements in the gray area, and Ti is uniformly distributed in the alloy. In addition, the minimum enthalpy of mixing (-7 KJ/mol) of Cr and Ta means that during the solidification process, Cr and Ta are more easily combined to form intermetallic compounds, which leads to the formation of Cr_2Ta in the Laves phase. It is evident from Figure 3e that the feathered Laves phase distribution forms a network with white dendritic areas. In addition, the presence of Cr was also detected in Figure 3a. As the laser cladding makes the 42-CrMo steel dilute, the Cr element in the substrate penetrates the coating.

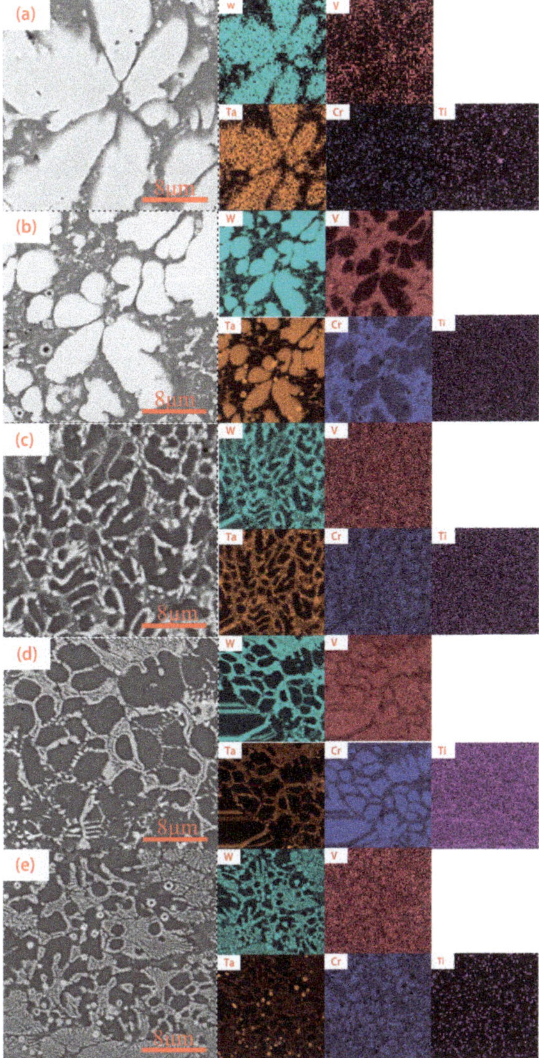

Figure 3. EDS of cross section of WVTaTiCr$_x$ RHEA coating: (**a**) Cr$_0$; (**b**) Cr$_{0.25}$; (**c**) Cr$_{0.5}$; (**d**) Cr$_{0.75}$; (**e**) Cr$_1$.

3.2. Phase Analysis

Phase analysis of five high-entropic alloy-coated samples with different contents of the Cr element was carried out, and the phase in the alloy was relatively simple. The main refractory metal elements are the BCC phase, so refractory high-entropy alloys are mostly simple BCC solid-solution phase, and sometimes the second phase is precipitated [35]. As shown in Figure 4, the positions of the diffraction peaks of the four components are approximately the same, consisting of the main peak of one (110) crystal plane and the low peak of (200) (211) crystal planes, mainly concentrated at the diffraction angles of 44.484°, 64.777° and 81.983°. Compared with the PDF card (PDF#34-0396), WVTaTiCr$_x$ RHEA coatings are based on the BCC solid-solution phase, and the addition of Cr promotes the precipitation of the Laves phase (Cr$_2$Ta), because the addition of Cr causes more severe lattice distortion in the alloy, thus opening a path for atomic diffusion [36]. At the same time, the enthalpy of mixing formed between Cr and other refractory elements is small, which reduces the enthalpy value in the alloy and the stability of the BCC phase, which leads to the precipitation of the Laves phase. From the thermodynamic point of view, if the enthalpy of mixing between the elements is positive, then the atoms of the two elements tend to repel each other, whereas the atoms of the two elements tend to attract each other to form a stable phase. The atomic radius of the Cr atom is the smallest and the easiest to diffuse, and the enthalpy of mixing between Cr and Ta is -7 KJ/mol. The binding ability between Cr and Ta is strong, and it is easier to form Cr_2Ta, which can be precipitated as the Laves phase. There are no other complex intermetallic compounds in the figure, so the high-entropy effect effectively inhibits the precipitation of intermetallic compounds.

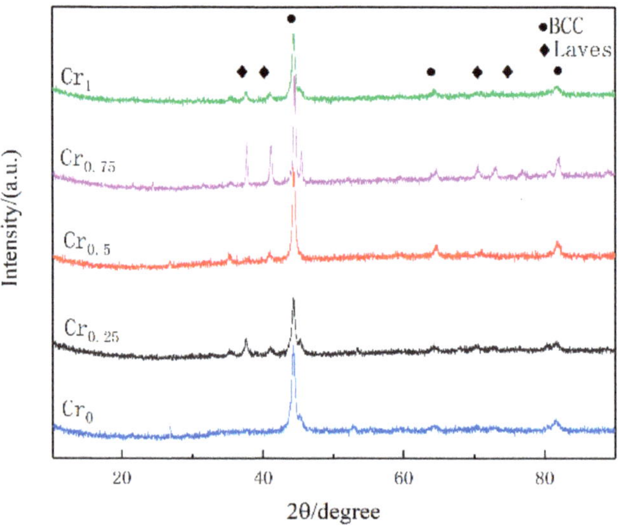

Figure 4. XRD of $WVTaTiCr_x$ high entropy alloy coating.

3.3. Hardness

Figure 5 summarizes the trend in hardness change of the cladding samples with different Cr contents. Figure 5a depicts the hardness distribution of the cross-section of high-entropy alloy coatings, and Figure 5b depicts the average hardness of high-entropic alloy coatings. As can be seen from Figure 5a, the entire cross-section is divided into three areas, namely, coating, heat-affected zone (HAZ), and substrate. The hardness of the whole coating area is significantly higher than that of the substrate. This is mainly because the laser cladding technology outputs a lot of heat momentarily through a high-energy laser beam, and the pre-set coating melts and solidifies rapidly, which produces a large supercooling degree during solidification, increases the nucleation rate, promotes a shorter growth time for each dendrite, collides with other grains, and grows again. The effect of grain refinement is significantly enhanced; thus, the hardness of the coating is generally higher than that of the substrate. As can be seen from Figure 5b, with the increase in Cr content, the average hardness of the coating also increases. According to previous studies, the hardness value of alloys has a strong relationship with the elements and phase composition. Combining XRD with SEM analysis, with the increase in Cr content, the Laves phase precipitation in the alloy also increases, and the precipitation strengthening effect is enhanced. In addition, with the increase in Cr content, the solid-solution strengthening effect in the alloy is enhanced, which further increases the hardness.

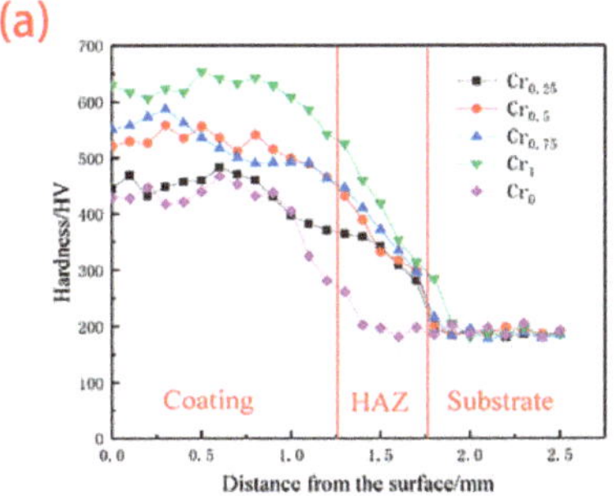
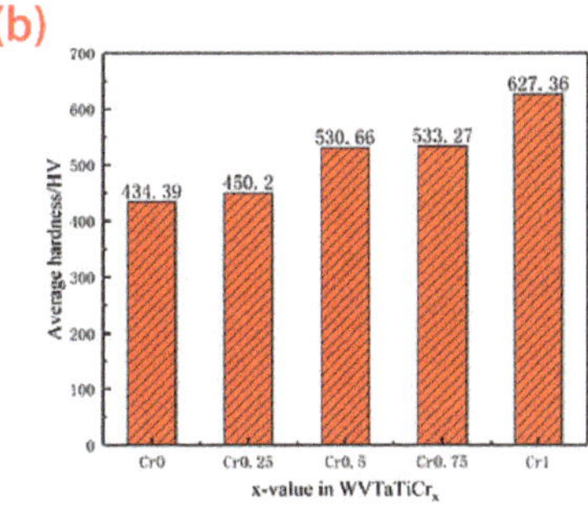

Figure 5. Hardness of WVTaTiCr$_x$ RHEA coating: (**a**) cross-section hardness; (**b**) average hardness of coating area.

3.4. Oxidation Resistance

Figure 6 summarizes the trend of oxide weight gain on different component coating surfaces after 50 h continuous oxidation of laser cladding samples at 800 °C. It can be seen from Figure 6 that, after adding the Cr element based on WVTaTi (Cr$_0$), the oxide weight gain decreases significantly. With the increase in oxidation time, the oxide weight-gain curve has continuity, and the content of Cr in the coating will affect the overall high-temperature oxidation resistance. In the first half of the experiment, the rate of oxidation weight increase is very fast, while, on the contrary, the rate of oxidation weight increase is slow after 20 h. Previous studies have shown that the high content of Cr in high-entropy alloys and the slow diffusion effect of high-entropy alloys can improve the high-temperature oxidation resistance of high-entropy alloys [12,37,38].

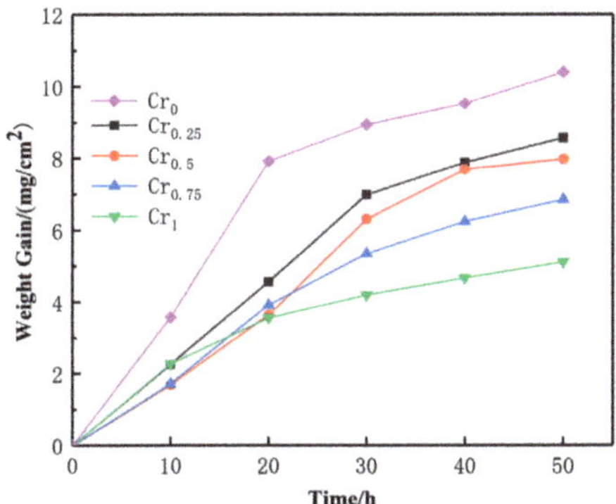

Figure 6. Weight increase of oxide on the surface of WVTaTiCr$_x$ RHEA coating.

Weight gain (mg/cm^2) and weight-gain rate (mg/(cm^2·h)) of WVTaTiCr$_x$ RHEA coating after oxidation at 800 °C for 50 h are shown in Table 3. When Cr is not added, the weight gain and oxidation rate of the WVTaTi high-entropy alloy coating oxide are 10.41 mg/cm^2 and 0.21 mg/(cm^2·h), respectively. With the increase in Cr content, the weight increases of oxide on the coating surface decreases. The addition of Cr resulted in the smallest increase in oxidation weight of the Cr$_1$ RHEA coating, with a mass increase of 5.12 mg/cm^2 and an oxidation rate of 0.10 mg/(cm^2·h). According to the oxidation-resistance rating standard of steel, superalloy, and high-temperature protective coatings (HB5258-2000) [39], it can be seen that Cr$_1$ RHEA coating belongs to the antioxidant alloys, and Cr$_{0.25}$, Cr$_{0.5}$, and Cr$_{0.75}$ belong to the sub-antioxidant alloys. In general, the proper addition of the Cr element can effectively improve the oxidation resistance of the coating.

Table 3. Weight gain (mg/cm^2) and weight-gain rate (mg/cm^2·h) of WVTaTiCr$_x$ RHEA coating after oxidation at 800 °C for 50 h.

Alloys	Cr$_0$	Cr$_{0.25}$	Cr$_{0.5}$	Cr$_{0.75}$	Cr$_1$
$\Delta m/s$	10.41	8.56	7.98	6.86	5.12
$\Delta m/(s{\cdot}t)$	0.21	0.17	0.16	0.14	0.10

Figure 7 shows the surface micromorphology of the coating of WVTaTiCr$_x$ RHEA after oxidation at 800 °C for 50 h. As can be seen from the figure, the coating surface generates large areas of irregular polygonal blocky oxides, which are overlapped with different sizes. To intuitively observe the distribution of elements, surface scanning analysis was performed on the selected areas of the oxidized surface, and the results are shown in Figure 8. From the figure, it can be seen that with the high-temperature oxidation in the same environment, the elemental distribution on the surface of the five coatings is approximately the same, and it can also be judged that the oxides on the surface are also basically similar. EDS showed that the oxygen content in the dendrite was greater than the interdendritic region, indicating that the oxidation resistance of the dendrites was weaker than the interdendritic region. V, Cr, Ti, and O are enriched elements, mainly concentrated in interdendritic regions; the proportion of W and Ta content on the entire surface is small. Moreover, as shown in Figure 3a, the dilution of laser cladding technology allows Cr from 42CrMo steel to diffuse into the WVTaTi coating. However, due to the small amount of Cr content, it cannot be detected after high-temperature oxidation.

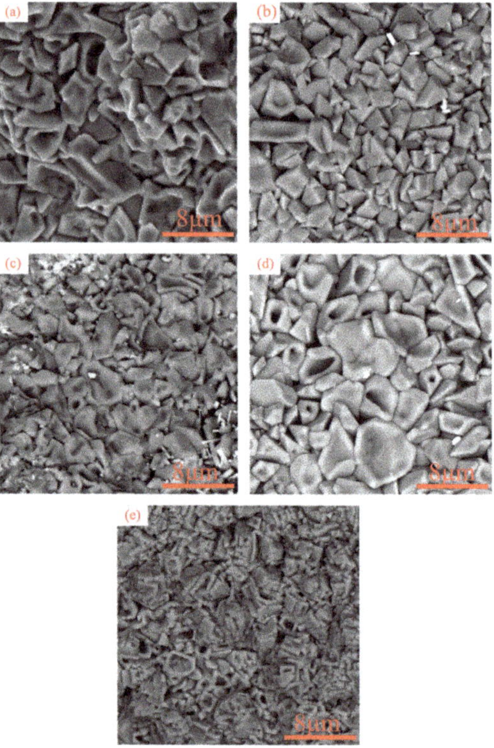

Figure 7. SEM of surface oxidation of WVTaTiCr$_x$ RHEA coating: (**a**) Cr$_0$; (**b**) Cr$_{0.25}$; (**c**) Cr$_{0.5}$; (**d**) Cr$_{0.75}$; (**e**) Cr$_1$.

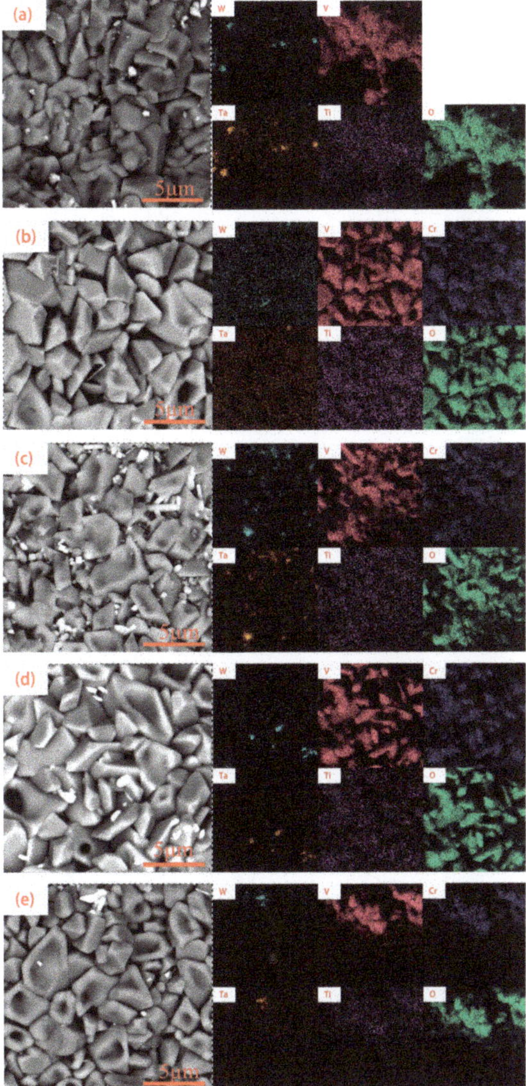

Figure 8. EDS element analysis of oxidation surface of WVTaTiCl$_x$ RHEA coating: (**a**) Cr$_0$; (**b**) Cr$_{0.25}$; (**c**) Cr$_{0.5}$; (**d**) Cr$_{0.75}$; (**e**) Cr$_1$.

To explore the types of oxides, XRD analysis was performed on the surface of the coatings after oxidation. As shown in Figure 9, the oxide mainly consists of CrTaO$_4$, TiO$_2$, and Cr$_2$O$_3$; the Cr$_1$ alloy has the highest oxide diffraction peak, indicating that more oxides are generated, making the oxide film on the coating surface uniform and dense. The dense oxide film inhibits the diffusion of oxygen atoms, reduces the oxidation rate, and improves oxidation resistance. This conclusion is also consistent with the results presented by the oxide weight-gain curve. The results show that adding Cr to the WVTaTi alloy can effectively improve high-temperature oxidation resistance.

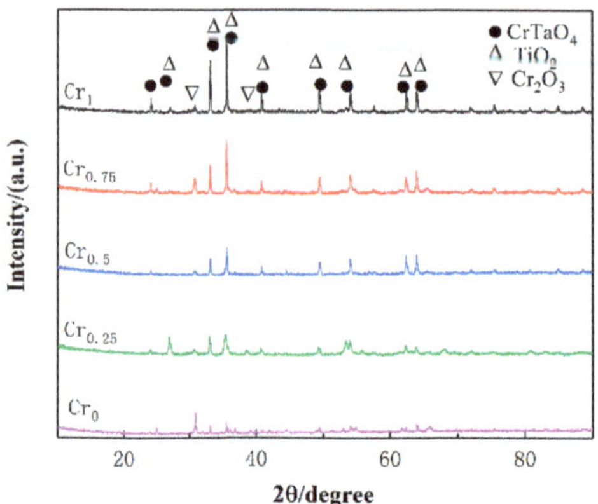

Figure 9. Surface oxide XRD of WVTaTiCr$_x$ RHEA coating.

Table 4 shows the standard Gibbs free energy of molar formation of the oxides of Cr, Ta, and Ti at 800 °C. Due to the smallest Gibbs free energy of TiO$_2$, it is first generated and exists stably. The Gibbs free energy of Ta$_2$O$_5$ is −598 KJ/mol, which is the second smallest, after the TiO$_2$. Due to the atomic radius of Ta being slightly larger than Cr and Ti, the diffusion rate is relatively slow. In addition, there is a lattice distortion effect inside high-entropy alloys, and atomic diffusion needs to overcome varying resistance. This slow diffusion effect makes it more difficult for Ta to generate oxides, so Ta$_2$O$_5$ does not appear on the surface. Although the standard Gibbs free energy of molar formation of Cr$_2$O$_3$ is greater than Ta$_2$O$_5$, the diffusion rate of Cr is faster than Ta, which also contributes to the formation of Cr$_2$O$_3$. Research shows that the Cr and Ta elements have a strong binding capacity with oxygen, so a small amount of CrTaO$_4$ can also be generated. With the addition of Cr, the internal lattice distortion effect of the WVTaTiCr alloy is the strongest, and the diffusion rate of each atom is the slowest, resulting in a decrease in the binding speed with external oxygen elements. This conclusion confirms the oxide XRD patterns and oxidation kinetic curves. The addition of Cr makes atomic diffusion slow, and the oxidation rate and weight-gain decrease, thereby improving the high-temperature oxidation resistance of the alloy.

Table 4. Standard Gibbs free energy of metal oxidation products of Cr, Ta, and Ti at 800 °C.

Oxide	Cr$_2$O$_3$	TiO$_2$	CrTaO$_4$	Ta$_2$O$_5$
ΔG (KJ/mol O$_2$)	−538	−713	−568	−598

3.5. Corrosion Resistance

The Tafel curves of each coating surface in 3.5 wt% NaCl solution are shown in Figure 10. Table 5 summarizes the corrosion potential (E$_{corr}$), corrosion current density (I$_{corr}$), and corrosion rate (V) in the Tafel curves of each coating. According to Figure 10, the self-corrosion potential of the WVTaTi (Cr$_0$) alloy is the lowest, and the self-corrosion potential of the alloy gradually increases with the addition of Cr. Among these, the WVTaCrTi (Cr$_1$) alloy coating has the highest corrosion potential. The data can be derived from Table 5; the corrosion potential and corrosion rate of WVTaTi (Cr$_0$) are −0.7783 V and 2.959 mm/a, respectively, and the corrosion potential and corrosion rate of the WVTaCrTi (Cr$_1$) alloy coating are −0.3198 V and 0.161 mm/a, respectively. The addition of Cr can promote the generation of the passive film on the surface of the alloy coating, while the oxide deposition of Cr can fill the vacancy of the passive film, making the surface passive film become denser and uniform, thus reducing the diffusion and penetration rate of ions in the NaCl solution, decreasing the corrosion rate and improving the corrosion resistance of the alloy surface coating.

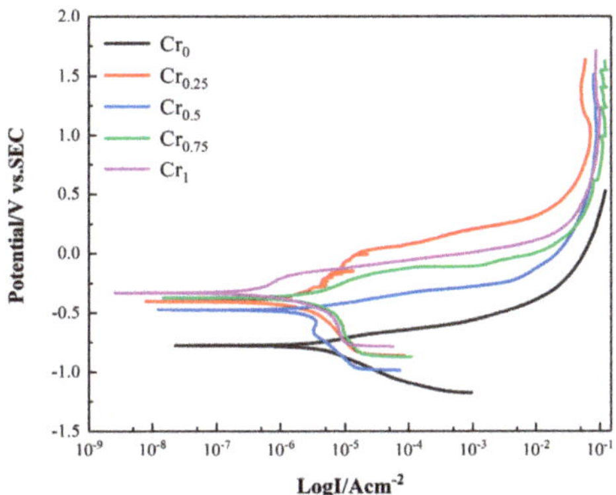

Figure 10. Tafel curve of WVTaTiCr$_x$ RHEA coating in 3.5%wt NaCl solution.

Table 5. Tafel Curve Electrochemical Parameters of WVTaTiCr$_x$ RHEA coating.

Alloys	E$_{corr}$ (V)	I$_{corr}$ (uA/cm^2)	V (mm/a)
Cr$_0$	−0.7783	6.466×10^{-6}	2.959
Cr$_{0.25}$	−0.4005	6.848×10^{-6}	2.545
Cr$_{0.5}$	−0.4663	1.059×10^{-6}	0.394
Cr$_{0.75}$	−0.3692	1.972×10^{-5}	7.329
Cr$_1$	−0.3198	4.337×10^{-7}	0.161

4. Conclusions

The WVTaTiCr$_x$ (x = 0, 0.25, 0.5, 0.75, 1) RHEA coating was prepared using laser cladding on the 42-CrMo steel surface. The effect of the Cr content on the microstructure and mechanical properties of the WVTaTiCr$_x$ (x = 0, 0.25, 0.5, 0.75, 1) alloy was studied. The analysis results are as follows:

(1) Laser-cladding-forming figures showed good metallurgical bonding between the coatings and the substrate, a uniform and dense coating structure, and a coating thickness of about 1 mm. The results indicate that laser cladding is feasible for the preparation of RHEA coatings.

(2) The WVTaTi (Cr$_0$) alloy coating consists of the BCC phase, and the addition of Cr promotes the precipitation of the Laves, which is uniformly distributed in the structure. The WVTaTiCr (Cr$_1$) exhibits superior mechanical properties, particularly in terms of its exceptional hardness, high-temperature oxidation resistance, and outstanding corrosion resistance.

(3) The successful preparation of WVTaTiCr alloy coating compensates for the performance shortcomings of some W-based binary alloys, providing a new idea for the development of heat insulation and radiation protection materials for nuclear reactors. The improved mechanical and oxidation properties of WVTaTiCr indicate the potential for future use in fusion applications.

Author Contributions: Conceptualization, Z.S. and Z.W.; experimental design, material preparation, and data collection, Z.X. and C.L.; writing—original draft preparation, Z.X.; writing—review and editing, Z.S.; formal analysis, Z.X.; investigation, C.L.; supervision, Z.S. and Z.W. All authors have read and agreed to the published version of the manuscript.

Funding: This study was supported by the Shandong Provincial Science and Technology Development Plan Project (grant number 2014GGX103009).

Institutional Review Board Statement: Not applicable.

Informed Consent Statement: Not applicable.

Data Availability Statement: The data presented in this study are available from the corresponding author upon reasonable request.

Conflicts of Interest: The authors declare no conflict of interest.

References

1. Senkov, O.N.; Wilks, G.B.; Miracle, D.B.; Chuang, C.P.; Liaw, P.K. Refractory high-entropy alloys. *Intermetallics* **2010**, *18*, 1758–1765. [CrossRef]
2. Kumar, R.; Torres, H.; Aydinyan, S.; Antonov, M.; Varga, M.; Hussainova, I.; Rodriguez Ripoll, M. Tribological behavior of Ni-based self-lubricating claddings containing sulfide of nickel, copper, or bismuth at temperatures up to 600 °C. *Surf. Coat. Technol.* **2023**, *456*, 129270. [CrossRef]
3. Hector, T.; Tugce, C.; Jens, H.; Janne, N.; Braham, P.; Manel, R.R. Tribological performance of iron- and nickel-base self-lubricating claddings containing metal sulfides at high temperature. *Friction* **2022**, *10*, 2069–2085. [CrossRef]
4. Couzinié, J.; Dirras, G.; Perrière, L.; Chauveau, T.; Leroy, E.; Champion, Y.; Guillot, I. Microstructure of a near-equimolar refractory high-entropy alloy. *Mater. Lett.* **2014**, *126*, 285–287. [CrossRef]
5. Juan, C.-C.; Tsai, M.-H.; Tsai, C.-W.; Lin, C.-M.; Wang, W.-R.; Yang, C.-C.; Chen, S.-K.; Lin, S.-J.; Yeh, J.-W. Enhanced mechanical properties of HfMoTaTiZr and HfMoNbTaTiZr refractory high-entropy alloys. *Intermetallics* **2015**, *62*, 76–83. [CrossRef]
6. Wu, Y.D.; Cai, Y.H.; Wang, T.; Si, J.J.; Zhu, J.; Wang, Y.D.; Hui, X.D. A Refractory Hf25Nb25Ti25Zr25 High-Entropy Alloy with Excellent Structural Stability and Tensile Properties. *Mater. Lett.* **2014**, *130*, 277–280. [CrossRef]
7. Yao, H.W.; Qiao, J.W.; Gao, M.C.; Hawk, J.A.; Ma, S.G.; Zhou, H.F.; Zhang, Y. NbTaV-(Ti, W) Refractory High-entropy Alloys: Experiments and Modeling. *Mater. Sci. Eng. A* **2016**, *674*, 203–211. [CrossRef]
8. Zhang, H.; Zhao, Y.; Cai, J.; Ji, S.; Geng, J.; Sun, X.; Li, D. High-strength NbMoTaX refractory high-entropy alloy with low stacking fault energy eutectic phase via laser additive manufacturing. *Mater. Des.* **2021**, *201*, 109462. [CrossRef]
9. Zhang, Y.; Yi-Wen, Z.; Feng-Ge, Z.; Tao, Y.; Chen, X.C. Effect of Powder Particle Size on Microstructure and Mechanical Property of Ni-Based P/M Superalloy Product. *J. Iron Steel Res.* **2003**, *10*, 71–74.
10. Ding, X.; He, J.; Zhong, J.; Wang, X.; Li, Z.; Tian, J.; Dai, P. Effect of Al Addition on Microstructure and Properties of CoCrNi Medium-Entropy Alloy Prepared by Powder Metallurgy. *Materials* **2022**, *15*, 9090. [CrossRef]
11. Thürlová, H.; Průša, F. Influence of the Al Content on the Properties of Mechanically Alloyed CoCrFeNiMn$_X$Al$_{20-X}$ High-Entropy Alloys. *Materials* **2022**, *15*, 7899. [CrossRef] [PubMed]
12. Kumar, T.S.; Sourav, A.; Murty, B.; Chelvane, A.; Thangaraju, S. Role of Al and Cr on cyclic oxidation behavior of AlCoCrFeNi$_2$ high entropy alloy. *J. Alloys Compd.* **2022**, *919*. [CrossRef]
13. Chen, L.; Zhou, Z.; Tan, Z.; He, D.; Bobzin, K.; Zhao, L.; Öte, M.; Königstein, T. High temperature oxidation behavior of Al$_{0.6}$CrFeCoNi and Al$_{0.6}$CrFeCoNiSi$_{0.3}$ high entropy alloys. *J. Alloys Compd.* **2018**, *764*, 845–852. [CrossRef]
14. Jiang, Y.; Zhao, N.; Peng, L.L.; Zhao, L.N.; Liu, M. Microstructure and mechanical properties at elevated temperatures of a new Al-containing refractory high-entropy alloy Nb-Mo-Cr-Ti-Al. *J. Alloys Compd.* **2016**, *195*, 104001. [CrossRef]
15. Qin, Y.; Liu, J.X.; Li, F.; Wei, X.; Wu, H.; Zhang, G.J. A high entropy silicide by reactive spark plasma sintering. *J. Adv. Ceram.* **2019**, *8*, 148–152. [CrossRef]
16. Yurchenko, N.; Stepanov, N.; Gridneva, A.; Mishunin, M.; Salishchev, G.; Zherebtsov, S. Effect of Cr and Zr on phase stability of refractory Al-Cr-Nb-Ti-V-Zr high-entropy alloys. *J. Alloys Compd.* **2018**, *757*, 403–414. [CrossRef]
17. Mooney, H.A.; Cropper, A.; Reid, W. Effects of grain size on the microstructure and texture of cold-rolled Ta-2.5W alloy. *Int. J. Refract. Met. Hard Mater.* **2016**, *58*, 25–136. [CrossRef]
18. Arshad, K.; Guo, W.; Wang, J.; Zhao, M.-Y.; Yuan, Y.; Zhang, Y.; Wang, B.; Zhou, Z.-J.; Lu, G.-H. Influence of vanadium precursor powder size on microstructures and properties of W–V alloy. *Int. J. Refract. Met. Hard Mater.* **2015**, *50*, 59–64. [CrossRef]
19. Dai, W.; Liang, S.; Luo, Y.; Yang, Q. Effect of W powders characteristics on the Ti-rich phase and properties of W–10 wt.% Ti alloy. *Int. J. Refract. Met. Hard Mater.* **2015**, *50*, 240–246. [CrossRef]
20. Ma, Y.; Han, Q.-F.; Zhou, Z.-Y.; Liu, Y.-L. First-principles investigation on mechanical behaviors of W–Cr/Ti binary alloys. *J. Nucl. Mater.* **2016**, *468*, 105–112. [CrossRef]
21. Rieth, M.; Dudarev, S.L.; De Vicente, S.G.; Aktaa, J.; Ahlgren, T.; Antusch, S.; Armstrong, D.E.J.; Balden, M.; Baluc, N.; Barthe, M.F.; et al. Recent progress in research on tungsten materials for nuclear fusion applications in Europe. *J. Nucl. Mater.* **2013**, *432*, 482–500. [CrossRef]
22. Xiao, X.; Liu, G.; Hu, B.; Wang, J.; Ma, W. Microstructure Stability of V and Ta Microalloyed 12%Cr Reduced Activation Ferrite/Martensite Steel during Long-term Aging at 650 °C. *J. Mater. Sci. Technol.* **2015**, *31*, 311–319. [CrossRef]
23. Senkov, O.N.; Wilks, G.B.; Scott, J.M.; Miracle, D.B. Mechanical properties of Nb$_{25}$Mo$_{25}$Ta$_{25}$W$_{25}$ and V$_{20}$Nb$_{20}$Mo$_{20}$Ta$_{20}$W$_{20}$ refractory high entropy alloys. *Intermetallics* **2011**, *19*, 698–706. [CrossRef]
24. Jin, X.; Liang, Y.; Bi, J.; Li, B. Non-monotonic variation of structural and tensile properties with Cr content in AlCoCr$_x$FeNi$_2$ high entropy alloys. *J. Alloys Compd.* **2019**, *798*, 243–248. [CrossRef]
25. Yan, X.; Guo, H.; Yang, W.; Pang, S.; Wang, Q.; Liu, Y.; Liaw, P.K.; Zhang, T. Al$_{0.3}$Cr$_x$FeCoNi high-entropy alloys with high corrosion resistance and good mechanical properties—ScienceDirect. *J. Alloys Compd.* **2021**, *860*, 158436. [CrossRef]

26. Ben, Q.; Zhang, Y.; Sun, L.; Wang, L.; Wang, Y.; Zhan, X. Wear and Corrosion Resistance of FeCoCr$_x$NiAl High-Entropy Alloy Coatings Fabricated by Laser Cladding on Q345 Welded Joint. *Metals* **2022**, *12*, 1428. [CrossRef]
27. Senkov, O.N.; Senkova, S.V.; Woodward, C.; Miracle, D.B. Low-density, refractory multi-principal element alloys of the Cr–Nb–Ti–V–Zr system: Microstructure and phase analysis. *Acta Mater.* **2013**, *61*, 1545–1557. [CrossRef]
28. Waseem, O.A.; Ryu, H.J. Powder Metallurgy Processing of a W$_x$TaTiVCr High-Entropy Alloy and Its Derivative Alloys for Fusion Material Applications. *Sci. Rep.* **2017**, *7*, 1926. [CrossRef]
29. Moshtaghi, M.; Safyari, M.; Mori, G. Hydrogen absorption rate and hydrogen diffusion in a ferritic steel coated with a micro- or nanostructured ZnNi coating. *Electrochem. Commun.* **2022**, *134*, 107169. [CrossRef]
30. Erario, M.d.l.Á.; Croce, E.; Moviglia Brandolino, M.T.; Moviglia, G.; Grangeat, A.M. A review on laser cladding of high-entropy alloys, their recent trends and potential applications. *J. Manuf. Process.* **2021**, *68*, 225–273.
31. Liu, H.; Gao, Q.; Dai, J.; Chen, P.; Gao, W.; Hao, J.; Yang, H. Microstructure and high-temperature wear behavior of CoCrFeNiWx high-entropy alloy coatings fabricated by laser cladding. *Tribol. Int.* **2022**, *172*, 107574. [CrossRef]
32. Zhou, J.; Kong, D. Friction–Wear performances and oxidation behaviors of Ti3AlC2 reinforced Co–based alloy coatings by laser cladding. *Surf. Coat. Technol.* **2021**, *408*, 126816. [CrossRef]
33. Qiu, X.-W.; Liu, C.-G. Microstructure and properties of Al2CrFeCoCuTiNix high-entropy alloys prepared by laser cladding. *J. Alloys Compd.* **2013**, *553*, 216–220. [CrossRef]
34. Zhou, J.-L.; Cheng, Y.-H.; Chen, Y.-X.; Liang, X.-B. Composition design and preparation process of refractory high-entropy alloys: A review. *Int. J. Refract. Met. Hard Mater.* **2022**, *105*, 105836. [CrossRef]
35. Senkov, O.N.; Scott, J.M.; Senkova, S.V.; Miracle, D.B.; Woodward, C.F. Microstructure and room temperature properties of a high-entropy TaNbHfZrTi alloy. *J. Alloys Compd.* **2011**, *509*, 6043–6048. [CrossRef]
36. Gao, X.; Chen, R.; Liu, T.; Fang, H.; Wang, L.; Su, Y. High deformation ability induced by phase transformation through adjusting Cr content in Co-Fe-Ni-Cr high entropy alloys. *J. Alloys Compd.* **2022**, *895*, 162564. [CrossRef]
37. Holcomb, G.R.; Tylczak, J.H.; Carney, C.S. Oxidation of CoCrFeMnNi High Entropy Alloys. *Jom* **2015**, *67*, 2326–2339. [CrossRef]
38. Zhang, Y.; Wu, H.; Yu, X.; Tang, D. Role of Cr in the High-Temperature Oxidation Behavior of Cr$_x$MnFeNi High-Entropy Alloys at 800 °C in Air. *Soc. Sci. Electron. Publ.* **2022**, *15*, 110211. [CrossRef]
39. Wang, H.; Tang, Q.; Li, X.; Dai, P. The effect of N on the oxidation resistance of CoCrFeMnNi high entropy alloy. *Met. Heat Treat.* **2017**, *42*, 7. (In Chinese) [CrossRef]

 *materials*

Article

Effect of Laser Beam Profile on Thermal Transfer, Fluid Flow and Solidification Parameters during Laser-Based Directed Energy Deposition of Inconel 718

Bo Chen [1,2,3], Yanhua Bian [1,3], Zhiyong Li [1,3,4,†], Binxin Dong [1,3], Shaoxia Li [1,3,4], Chongxin Tian [1,3,4], Xiuli He [1,3,4,*] and Gang Yu [1,2,3,4,*]

[1]   Institute of Mechanics, Chinese Academy of Sciences, Beijing 100190, China; chenbo@imech.ac.cn (B.C.); bianyanhua@imech.ac.cn (Y.B.); lizhiyong@imech.ac.cn (Z.L.); dongbinxin@imech.ac.cn (B.D.); lisx@imech.ac.cn (S.L.); tianchongxin@imech.ac.cn (C.T.)

[2]   Center of Materials Science and Optoelectronics Engineering, University of Chinese Academy of Sciences, Beijing 100190, China

[3]   School of Engineering Science, University of Chinese Academy of Sciences, Beijing 100190, China

[4]   Guangdong Aerospace Research Academy, Guangzhou 511458, China

*   Correspondence: xlhe@imech.ac.cn (X.H.); gyu@imech.ac.cn (G.Y.)

†   Current address: Welding and Additive Manufacturing Centre, Cranfield University, Cranfield MK43 0AL, Bedfordshire, UK.

Abstract: The profile of the laser beam plays a significant role in determining the heat input on the deposition surface, further affecting the molten pool dynamics during laser-based directed energy deposition. The evolution of molten pool under two types of laser beam, super-Gaussian beam (SGB) and Gaussian beam (GB), was simulated using a three-dimensional numerical model. Two basic physical processes, the laser–powder interaction and the molten pool dynamics, were considered in the model. The deposition surface of the molten pool was calculated using the Arbitrary Lagrangian Eulerian moving mesh approach. Several dimensionless numbers were used to explain the underlying physical phenomena under different laser beams. Moreover, the solidification parameters were calculated using the thermal history at the solidification front. It is found that the peak temperature and liquid velocity in the molten pool under the SGB case were lower compared with those for the GB case. Dimensionless numbers analysis indicated that the fluid flow played a more pronounced role in heat transfer compared to conduction, especially in the GB case. The cooling rate was higher for the SGB case, indicating that the grain size could be finer compared with that for the GB case. Finally, the reliability of the numerical simulation was verified by comparing the computed and experimental clad geometry. The work provides a theoretical basis for understanding the thermal behavior and solidification characteristics under different laser input profile during directed energy deposition.

Keywords: laser profile; thermal behavior; fluid flow; dimensionless number; cooling rate

Citation: Chen, B.; Bian, Y.; Li, Z.; Dong, B.; Li, S.; Tian, C.; He, X.; Yu, G. Effect of Laser Beam Profile on Thermal Transfer, Fluid Flow and Solidification Parameters during Laser-Based Directed Energy Deposition of Inconel 718. *Materials* **2023**, *16*, 4221. https://doi.org/10.3390/ma16124221

Academic Editor: Jun Liu

Received: 13 April 2023
Revised: 27 May 2023
Accepted: 5 June 2023
Published: 7 June 2023

1. Introduction

Additive manufacturing (AM) technology is increasingly becoming a popular manufacturing technique for producing parts from various metals and alloys, such as stainless steels, Ni-based superalloys, Al-alloys, etc. [1–4]. Among AM techniques, laser-based directed energy deposition (L-DED) is considered as a promising manufacturing method due to its unique merits, such as little heat affected zone, low dilution, and fine grain size [5,6]. It is widely used in material manufacturing, such as damaged parts repairing [7], metallic coating [8], etc. During the process of L-DED, the metallic substrate absorbs the energy of the laser beam and forms a small molten pool where powder particles are transported through a powder delivery system. As the laser moves along together with the feeding system, the solidified clad layer forms. Complex physical phenomena, such as laser–powder interaction, mass addition, fluid flow and temperature changes, are usually coupled in the process [9]. Due to the highly transient process of rapid heating and cooling, experimental measurement of those complex physical phenomena is difficult. As a result, an accurate numerical model based on the physics, validated using the experimental data, can help us understand the underlying theory of molten pool dynamics during L-DED processing.

A substantial amount of numerical model has been proposed based on reasonable simplifications because of the complex physical phenomena involved in the L-DED process. Dortkasli et al. [10] developed an efficient finite element model for thermal process to predict the characteristics of the clad layer in the L-DED process. However, the fluid flow of the liquid metal was not taken into account in the simulation. Sun et al. [11] established a 3D numerical model to study the mass and thermal transport in a high deposition rate laser cladding process. In this model, the interaction between powder particles and the molten pool surface was considered. Gan et al. [12] proposed a self-consistent model, which incorporated the fluid dynamics and multicomponent mass transfer. In this model, mass transfer and fluid flow were coupled successfully and the solidification parameters were also analyzed. Li et al. [13] developed a multi-physics model for DED process of functionally gradient materials (FGMs) fabrication. This model can predict the solute distribution and geometry of the manufactured FGMs. However, none of the above models took the interaction between laser beam and powder flow into account directly.

The attenuation of laser beam by powder particles and the heat absorption of powder particles are two key factors which dominate the energy input and further influence the fluid flow throughout the L-DED process. Bayat et al. [14] established a multi-physics numerical model to investigate the molten pool variation affected by different particle velocities. In this model, the impingement velocity and temperature rise of powder particles were taken into account. However, the model did not contain the influence of powder streams on the laser intensity. Song et al. [15] created a numerical model for laser metal deposition process and used various process parameters to compare with experimental results. In this model, the attenuation of laser was calculated using a semi-empirical model. However, the powder temperature distribution on the deposition surface was not involved in this model. Wang et al. [16] developed a 3D numerical model for L-DED, considering thermal-fluid transport and laser–powder interaction. Although the effect of laser–powder interaction was considered, thermal convection and radiation of the powders were ignored. Moreover, no detailed consequence of the attenuated laser intensity and powder temperature rise were reported. Wu et al. [17] fabricated a 2D numerical model to analyze the thermal-fluid transport inside the molten pool. In this model, whole-phase laser–powder coupling and material deposition were comprehensively considered. Yet, the analysis of the L-DED process is limited by the model dimension.

The process parameters determine the solidification conditions and further the quality of the molten pool. Intensity profile of laser beam is a key parameter which plays a significant role in the laser material processing. Gaussian, super-Gaussian and flat top profile are three typical distribution of laser beam intensity. For a focused fiber laser beam, the beam profile transforms from flat-top to Gaussian from the focal plane to the far field [18]. The super-Gaussian function is used to describe the spatial profile of laser beam from Gaussian profile to a flat-top profile. Several researches [18–23] have been proposed to explore the effect of laser intensity on the laser-induced molten pool. Han et al. [19] used different types of laser beam modes to study their influence on the molten pool and concluded that the molten pool under Gaussian beam had the greatest depth. Ayoola et al. [20] claimed that a Gaussian laser beam could result in deeper weld pool compared with that for a laser beam with top-hat profile. Kaplan [18] indicated that a top-hat beam and Gaussian beam would cause different, steep keyhole shapes during deep penetration laser welding. Huang et al. [21] found that a top-hat laser beam was more suitable for manufacturing a denser prototype compared with a Gaussian beam in laser powder bed fusion (LPBF). This study showed that a lower aspect ratio and less keyhole porosities for the clad were observed in the case using a top-hat laser beam. Yuan et al. [22] investigated the thermal-fluid transport inside the molten pool under the laser beams with flat-top and Gaussian profile in LPBF. The results showed that these two laser beams had significantly different effect on the clad appearance. Moreover, the flat-top laser showed great potential for controlling the directed growth of grains. Wu et al. [23] compared the different phenomenon of thermal behavior and fluid flow inside the molten pool under Gaussian and super-Gaussian beam via a two-dimension model for the L-DED process. It was found that the beam with super-Gaussian profile had less influence on the depth of the molten pool and had better stability than that with the Gaussian distribution.

As seen from the above discussions, powder addition and laser intensity are influential on mass input and heat flux input applied to the deposited surface, which would largely affect the thermal-fluid transport. The research on mutual coupling influence of the laser energy input and powder addition behavior is essential for a comprehensive understanding of the L-DED process. There have been some studies [14–16] on the mutual coupling effect of powder addition behavior and the laser intensity input with Gaussian distribution. However, there has been few researches on the laser–powder interaction and corresponding heat transport of powders under the laser intensity input with super-Gaussian distribution. Therefore, in this paper, an improved thermal-fluid model including a laser–powder interaction model and a metal deposition model was established to explore the thermal-fluid transport and solidification characteristics under two types of laser beams (Gaussian and super-Gaussian) during single-track L-DED process of Inconel 718. The moving mesh method [12,15] was used to capture liquid/gas interface dynamically and the apparent heat capacity method [16,17] was used to track the solid/liquid interface. Subsequently, several dimensionless numbers were analyzed to explain the molten pool evolution during the L-DED process using the calculated results. Furthermore, the solidification parameters which are obtained from thermal history were used to reveal the solidification patterns and grain morphology of the molten pool. Finally, the reliability of the model was verified by comparing the morphology of the molten pool from calculations and experiments.

2. Experimental Procedure

To validate the consequence of the 3D transient thermal-fluid model, single-track deposition experiments were carried out on a fiber laser system equipped with a coaxial powder feeding nozzle. The schematic diagram and experimental setup for coaxial L-DED process are plotted in Figure 1.

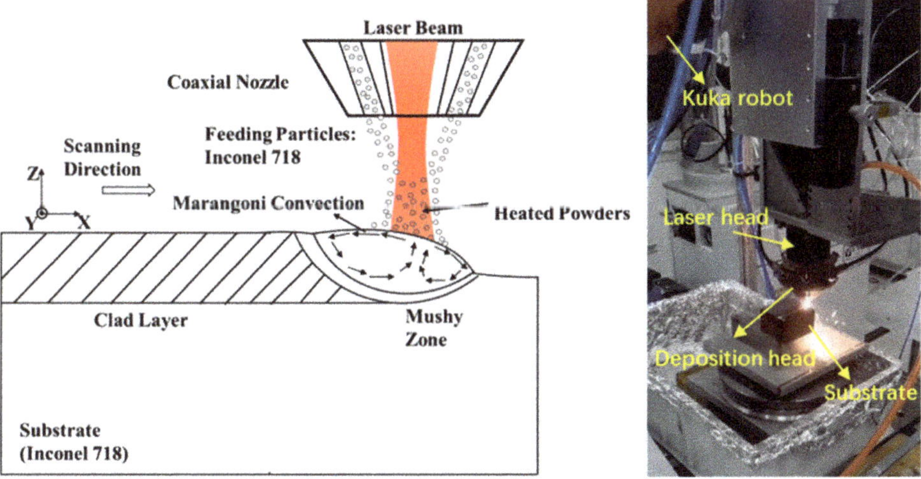

Figure 1. Schematic diagram and experimental setup for coaxial L-DED process.

The substrate supplied by Shanghai Linzhi Metal Materials Co., Ltd. (Shanghai, China) and powder supplied by Jiangsu Vilory Advanced Materials Technology Co., Ltd. (Xuzhou, China) were both Inconel 718. The substrate with the dimensions of 100 mm × 40 mm × 40 mm was sanded using 40 #, 180 # and 400 # silicon carbide sandpaper and washed with ethanol before experiments. Powder particle diameter is in the range of 59–123 μm, as shown in Figure 2a, and the cumulative distribution is plotted in Figure 2b.

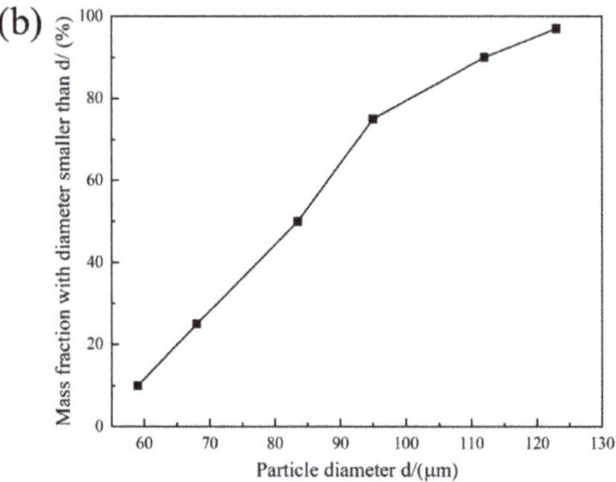

Figure 2. (**a**) SEM image and (**b**) Cumulative distribution of Inconel 718 powder particles.

The wavelength of the fiber laser (YLS-10000-CUT, IPG Photonics Corporation, Oxford, MA, USA) used in the experiments was 1070 nm. The spot radius of laser beam was set as 1.6 mm. Both the carrier gas and the protective gas were argon with a purity of 99%. A continuous coaxial powder feeding nozzle with two channels for shielding gas and an annular wedge-shaped channel for powder and carrying gas was employed to provide a protective atmosphere and the added powder material. The detailed structure of the continuous coaxial powder feeding nozzle can be checked in our previously published paper [24]. The movement of the laser cladding system was implemented and controlled by a six-axis KUKA robot. More parameters used for experiments are presented in detail in Table 1.

Table 1. Processing parameters used in L-DED experiments.

Processing Parameters	Laser Spot Radius (mm)	Powder Feeding Rate (g/min)	Laser Power (W)	Scanning Speed (mm/s)
value	1.6	5	1500–2100	10

3. Mathematical Model

As stated in the Introduction section, L-DED of alloys is a very complex process. In order to establish a model for the whole L-DED process, some reasonable assumptions are as follows [9,25,26]:

(1) Liquid metal flow inside the molten pool is considered as Newtonian, laminar and incompressible.
(2) The vaporization is ignored because the calculated peak temperature of the system is lower than the boiling point of the material.
(3) The mushy zone whose temperature is between the solidus and liquidus is described using a porous medium with isotropic permeability.
(4) The impact of shielding gas is not taken into consideration because of its minimal pressure.

(5) The powder instantly melts when it gets in touch with the deposition surface and is treated as a continuous and uniform phase.

3.1. Governing Equations

According to the above assumptions, the conservation equation of mass is:

$$\frac{\partial \rho}{\partial t} + \nabla \cdot (\rho \boldsymbol{u}) = 0 \tag{1}$$

where, $\boldsymbol{u}$ and ρ denote the velocity and density of metal fluid, respectively.

The conservation equation of momentum is:

$$\frac{\partial(\rho \boldsymbol{u})}{\partial t} + \nabla(\rho \boldsymbol{u}\boldsymbol{u}) = -\nabla p + \nabla(\mu(\nabla \boldsymbol{u} + \nabla \boldsymbol{u}^T)) + \boldsymbol{F_b} + \boldsymbol{F_m} \tag{2}$$

where, p is the pressure, and μ is the dynamic viscosity. The second term on the right hand of Equation (2) is related to viscosity shear stress. $\boldsymbol{F_b}$ is used to describe the force of buoyancy using Boussinesq approximation [27], and its formula is given as,

$$\boldsymbol{F_b} = (1 - \beta(T - T_{ref}))\rho \boldsymbol{g} \tag{3}$$

where, $\boldsymbol{g}$ is the gravity acceleration, β is the coefficient of volumetric expansion, and T_{ref} is the reference temperature. The last term on the right hand of Equation (2) represents the damping force and it is given as [28],

$$\boldsymbol{F_m} = -A_{mushy}\frac{(1 - f_l)^2}{f_l^3 + \varepsilon}\boldsymbol{u} \tag{4}$$

where, A_{mushy} is a huge constant (10^4 in this paper) related to the damping forces, and ε is a positive constant (10^{-3} in this paper) to avoid the division by zero. f_l represents the fraction of liquid phase in the mushy zone, which could be given as:

$$f_l = \begin{cases} 0 & (0 < T < T_s) \\ \frac{T - T_s}{T_l - T_s} & (T_s < T < T_l) \\ 1 & T > T_l \end{cases} \tag{5}$$

where, T is the local temperature of the system. T_s and T_l denote the solidus and liquidus temperature of the liquid metal, respectively.

The conservation equation of thermal energy is:

$$\rho c_p \frac{\partial T}{\partial t} + \rho c_p \boldsymbol{u} \cdot \nabla T = k \nabla^2 T \tag{6}$$

where, c_p and k represent the heat capacity and thermal conductivity of the metal, respectively.

3.2. Heat Source Model

The laser heat source applied on the liquid/gas interface is assumed to be super-Gaussian beam (SGB) and its spatial laser intensity profile can be expressed as [29]:

$$I_{SGB}(x, y) = \frac{4^{1/N}NP}{2\pi r_l^2 \Gamma(N/2)} \exp[-2(\frac{(x - v_s t)^2 + y^2}{r_l^2})^{N/2}] \tag{7}$$

where, P is the input laser power and N represents the order of super-Gaussian. Γ denotes the gamma function and $\Gamma(N/2)$ is the SGB power distribution. The beam radius is related to the axial propagation of the laser and it is given as [18]:

$$r_l = r_0\sqrt{1 + (\frac{z - z_0}{z_R})^2}$$

(8)

where, z_R is the Rayleigh range. While $N = 1$, there is a Gaussian profile for the energy source, which is presented in Figure 3a. According to the high-power fiber laser used in laser cladding experiments, the value of N is approximately 5. As shown in Figure 3b, the heat source intensity looks like a top-hat and the total energy is divided equally within the limited region.

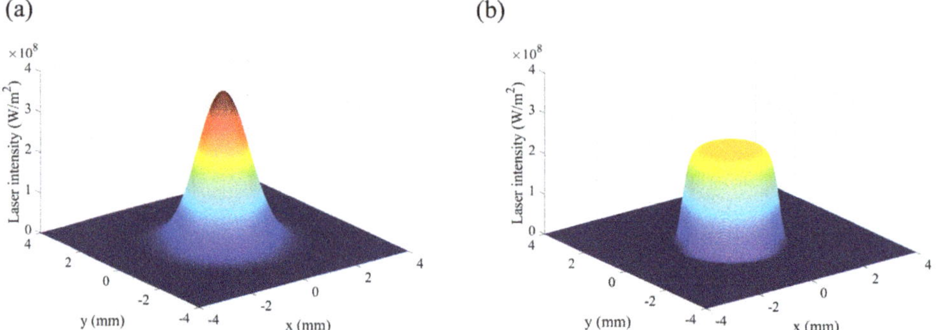

Figure 3. Laser intensity under different super-Gaussian order of (**a**) 1, (**b**) 5 (laser power: 1500 W; laser spot: 1.6 mm).

3.3. Laser–Powder Interaction

For laser–powder interaction in the L-DED process, two main physical phenomena are involved: laser intensity attenuation and temperature rise of powders. The concentration distribution of the coaxial powder stream is assumed to be a Gaussian distribution [30] and it is given as:

$$N(r, z) = \frac{2M_p}{m_p \pi v_p R_p^2(z)} \exp\left(-\frac{2 \cdot r^2}{R_p^2(z)}\right)$$

(9)

where, M_p is the mass flow rate, R_p represents the equivalent Gaussian radius at each cross section, v_p denotes the powder velocity and m_p is the mass weight of a single powder particle.

Due to the absorption and scattering of the powder stream, the attenuation of the laser beam intensity can be calculated using Beer Lambert law [31]:

$$dI = -Q_{ext}\pi r_p^2 I(r, z)N(r, z)dz$$

(10)

where, Q_{ext} is the extinction coefficient. In this study, the average particle radius $r_p = 42\ \mu m$, and laser wavelength $\lambda_l = 1.07\ \mu m$. The laser wavelength is much smaller than the particle size. Thus, it is assumed that the powder particles absorb most of the attenuated energy and the extinction coefficient $Q_{ext} = 1$ [31].

By integrating Equation (10) along the z-direction, the attenuated laser beam intensity $I_a\ (r, z_k)$ can be calculated using the following formula:

$$I_a(r, z_k) = \begin{cases} I_{SGB}(r, z_0) & k = 0 \\ I_a(r, z_{k-1})\exp(-Q_{ext}\pi r_p^2 N(r, z_{k-1})(z_k - z_{k-1})) & k \geq 1 \end{cases}$$

(11)

where, z_0 is the initial interaction position of the laser beam and powder stream.

The powder particle temperature starts to rise as it absorbs the laser energy. Moreover, as it absorbs sufficient energy, phase changes could occur before reaching the liquid/gas interface. The temperature rise for a single powder particle follows the heat balance equation [17]:

$$m_p c_p \frac{dT_p}{dt} = \eta_p I_a(r,z) \pi r_p^{2} - h_p A_p (T_p - T_\infty) - \varepsilon \sigma A_p (T_p^4 - T_\infty{}^4) - m_p L_m \frac{df}{dt} \tag{12}$$

where, η_p denotes the laser absorptivity, h_p is convective heat transfer coefficient of the powder, A_p is the surface area of the particle, T_∞ represents the temperature of surrounding gas, and L_m represents the latent heat of the particle material. The first term on the right hand of Equation (12) represents the heat absorbed by powder particles. The second and third term denote the convection and radiation between powder particles and the environment, respectively. The last term is relative to the phase change of powder particles.

3.4. Phase Change

After the temperature of the heated metal exceeds the liquidus, a solid–liquid mixed zone begins to appear. The thermal property of the mixed-region between solid and liquid is determined using a linear equivalent treatment, the formula is as follows:

$$\rho = \theta_s \rho_s + \theta_l \rho_l \tag{13}$$

$$k = \theta_s k_s + \theta_l k_l \tag{14}$$

where, θ_i, ρ_i and k_i represent the volume fraction, density and thermal conduction of i phase, respectively. The phase change in the molten pool is tracked using the apparent heat capacity method and the related formula is expressed as [32]:

$$c_p = \frac{1}{\rho}(\theta_s \rho_s c_{ps} + \theta_l \rho_l c_{pl}) + L_m \frac{\partial \alpha_m}{\partial T} \tag{15}$$

where, α_m is a distribution function of the latent heat which can be given as:

$$\alpha_m = \frac{1}{2} \frac{\theta_l \rho_l - \theta_s \rho_s}{\rho} \tag{16}$$

The thermal property of the material applied in the calculation is presented in Table 2.

Table 2. Material properties of Inconel 718 applied in calculation [33].

Property	Value	Unit
Density of solid, ρ_s	7676	kg/m^3
Density of liquid, ρ_l	7400	kg/m^3
Liquid viscosity, μ	0.006	Pa·s
Solid temperature, T_s	1533	K
Liquid temperature, T_l	1609	K
Latent heat of fusion, L_f	2.09×10^5	J/kg
Thermal conductivity, k	$0.5603 + 0.0294T - 7.0 \times 10^{-6}\,T^2$	W/(m·K)
Specific heat of solid, $C_{p,s}$	625	J/(kg·K)
Specific heat of liquid, $C_{p,l}$	725	J/(kg·K)
Surface tension coefficient, γ	-0.00011	N/(m·K)
Thermal expansion coefficient, β	1.63×10^{-5}	1/K
Stefan-Boltzmann constant, σ	5.67×10^{-8}	W/(m^2·K^4)
Equivalent emissivity, ε	0.5	-

3.5. Free Surface Tracking

The free surface is tracked explicitly using the Arbitrary Lagrange–Euler (*ALE*) method and its normal velocity can be expressed as [12]:

$$V_{L/G} = \boldsymbol{u} \cdot \boldsymbol{n} + \boldsymbol{v_a} \cdot \boldsymbol{n} \tag{17}$$

where, $\boldsymbol{n}$ stands for the unit normal vector of the free surface and $\boldsymbol{v_a}$ is the related moving velocity due to mass addition, which can be described as:

$$\boldsymbol{v_a} = \frac{2\eta_c M_p}{\rho_p \pi R_p{}^2} \exp\left[-2\frac{(x - v_s t)^2 + y^2}{R_p{}^2}\right]\phi_s(T)\boldsymbol{e_z} \tag{18}$$

where, $\phi_s(T)$ is a smooth function, which is defined as:

$$\phi_s(T) = \begin{cases} 0 & T < T_s \\ 0.5 + 0.5\sin(\pi\frac{T-(T_l+T_s)/2}{T_l-T_s}) & T_s < T < T_l \\ 1 & T > T_l \end{cases} \tag{19}$$

where, η_c is the capture efficiency of the powders. The smooth function is used to ensure that the powder could only be captured by the melt metal. On the other hand, the powder has no influence on the no-melting substrate whose temperature is below the melting point.

3.6. Boundary Conditions

Due to powder injection, the free surface grows into a curved surface which separates the gas and liquid phases. Based on the previous assumption that as long as powders attach to the molten pool, they melt instantly and mix with the liquid metal, the free surface is treated as continuous media. Only half of the calculated domain is contained in this model because of the symmetry during the L-DED process.

3.6.1. Momentum Boundary Conditions

The liquid/gas interface stress tensor can be divided into two temperature-dependent parts in normal and tangential direction, which can be given as [7]:

$$\boldsymbol{F}_{L/G} = \sigma\kappa\boldsymbol{n} - \gamma(\nabla T - (\nabla T \cdot \boldsymbol{n})\boldsymbol{n}) \tag{20}$$

The first force is capillary force in normal direction pointing into the liquid metal. κ represents the curvature of the free surface and σ denotes the surface tension, which are expressed as follows [26]:

$$\kappa = \nabla \cdot \boldsymbol{n} \tag{21}$$

$$\sigma = \sigma_{ref} + \gamma(T - T_{ref}) \tag{22}$$

where, T_{ref} and σ_{ref} represent the reference temperature and surface tension, respectively. The second force of Equation (20) is the Marangoni force in the tangential direction which is related to variation of the surface tension. γ denotes the surface tension coefficient.

Due to the symmetry of the surface conditions, the fluid velocity is limited on the symmetry surface, which means that y-direction flow is forbidden in the plane, thus, the law of limited fluid flow is given as:

$$\boldsymbol{u} \cdot \boldsymbol{n}_{sp} = 0 \tag{23}$$

where, $\boldsymbol{n}_{sp}$ is the normal phasor of the plane.

Accordingly, the stress boundary conditions in the symmetric plane can be expressed as:

$$\left\{\mu[\nabla\boldsymbol{u} + (\nabla\boldsymbol{u})^T]\right\}\boldsymbol{n}_{sp} - ((\left\{\mu[\nabla\boldsymbol{u} + (\nabla\boldsymbol{u})^T]\right\}\boldsymbol{n}_{sp}) \cdot \boldsymbol{n}_{sp})\boldsymbol{n}_{sp} = 0 \tag{24}$$

3.6.2. Thermal Boundary Conditions

The heat flux applied on the free surface can be expressed as:

$$-k\frac{\partial T}{\partial n} = \eta_l I_a - q_p - h_c(T - T_\infty) - \varepsilon\sigma(T^4 - T_\infty{}^4) \tag{25}$$

where, I_a represents the attenuated laser power intensity which is calculated using Equation (11). η_l is the laser absorptance of the substrate. q_p is a source term carried by the heated powders and it can be expressed as:

$$q_p = \begin{cases} M_p''\left[L_m + c_{p,s}(T_s - T_p) + c_{p,l}(T - T_l)\right]\phi_s(T) & T_p < T_s \\ M_p''\left[(1 - f_l)L_m + c_{p,l}(T - T_l)\right]\phi_s(T) & T_s < T_p < T_l \\ M_p'' c_{p,l}(T - T_p)\phi_s(T) & T_p > T_l \end{cases} \tag{26}$$

where, T_p is the powder temperature on the liquid/gas interface which is also obtained from the laser–powder couple model. M_p'' is the mass flux due to powder addition and it is given as follows:

$$M_p'' = \frac{2\eta_c M_p}{\pi R_p{}^2}\exp\left[-2\frac{(x - v_s t)^2 + y^2}{R_p{}^2}\right] \tag{27}$$

The third and fourth term of Equation (25) are the heat loss due to the thermal convection and environment radiation, respectively.

The symmetric plane is an adiabatic boundary, which indicates that heat flux could not pass through the plane. The formula of the adiabatic boundary condition is given as:

$$k\nabla T \cdot \boldsymbol{n}_{sp} = 0 \tag{28}$$

The bottom and side surfaces are considered as imaginary surfaces, for which the heat convection is employed as [34]:

$$-k\frac{\partial T}{\partial n} = -h_{cs}(T - T_\infty) \tag{29}$$

where, h_{cs} represents the heat transfer coefficient, with an estimate value of 1250 W/(m^2·K), based on the findings of Khandkar et al. [35].

3.7. Numerical Procedure

The three-dimensional heat transfer model was used to conduct nonlinear transient simulation based on the commercial software COMSOL v5.4 Multiphysics. The model was built using three-dimensional Cartesian coordinate. The positive x-axis direction was consistent with the laser moving direction, the y-axis was the cross-section direction and the positive z-axis was the deposition direction. The domain with the dimension of 9 mm × 3 mm × 2 mm (x × y × z) was divided into free tetrahedron mesh, as shown in Figure 4. To calculate the temperature field, fluid dynamics and free surface moving of the molten pool accurately, finer meshes were employed near the laser scanning path. The corresponding minimum and maximum grid space were 40 μm and 100 μm, respectively, and the number of grids was about 200,000. Moreover, a time-dependent solver with adaptive time stepping was used in the model.

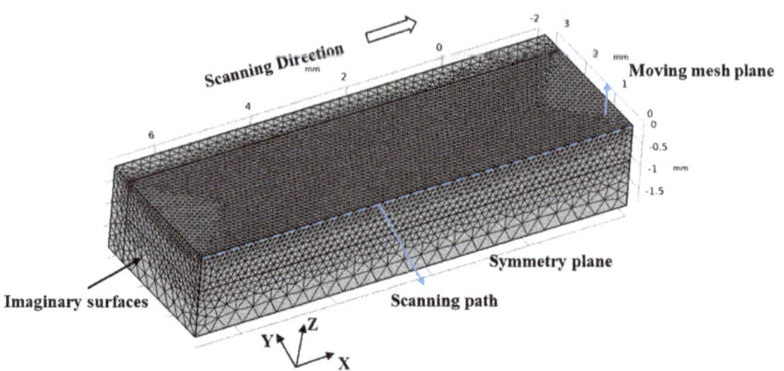

Figure 4. Schematic illustrations of computational domain.

4. Results and Discussion

4.1. Heat Transport of Powders

As the attenuated laser beam reaches the liquid/gas interface, the total intensity loss was calculated from the sum of the intensity loss along z direction, and the formula can be given as:

$$I_{loss}(r,z) = \sum_{z=z_0}^{z} I_a(r,z)[1 - \exp(-Q_{ext}\pi r_p^2 N(r,z)\Delta z)] \tag{30}$$

where, I_a represents the attenuated laser intensity. Figure 5a,b depict the total intensity loss (TIL) when the laser reaches the deposition surface. The intensity loss has a peak value in the central region, which is attributed to the higher powder stream concentration. The peak value of the TIL for the SGB case is approximately 37.3% lower than that for the GB case and the corresponding maximum values are 3.54×10^6 W/m^2 and 5.65×10^6 W/(m·K), respectively.

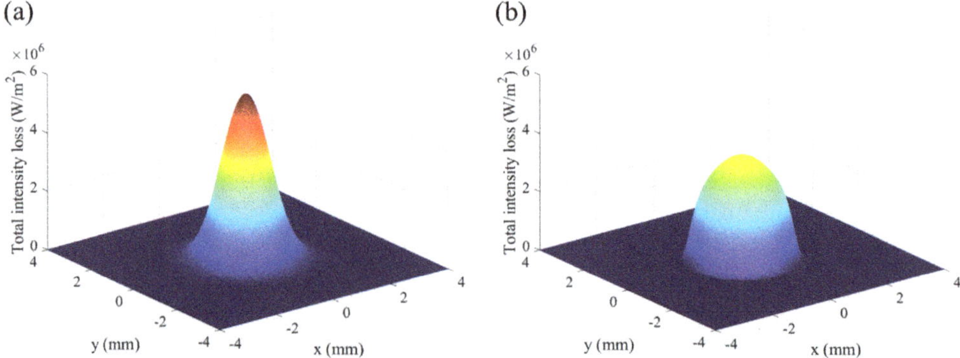

Figure 5. Total intensity loss after the laser reaches deposition surface under (**a**) GB, (**b**) SGB. (laser power: 2100 W; powder feeding rate: 5 g/min; laser spot: 1.6 mm; scanning speed: 10 mm/s).

Total power loss (TPL) after laser passes through the powder stream can be calculated by integrating the TIL over the entire deposition surface and the corresponding results are shown in Figure 6. It is found that TPL grows linearly with the increase in laser power for both cases and the TPL for the SGB case is approximately 2.0% higher than that for the GB case.

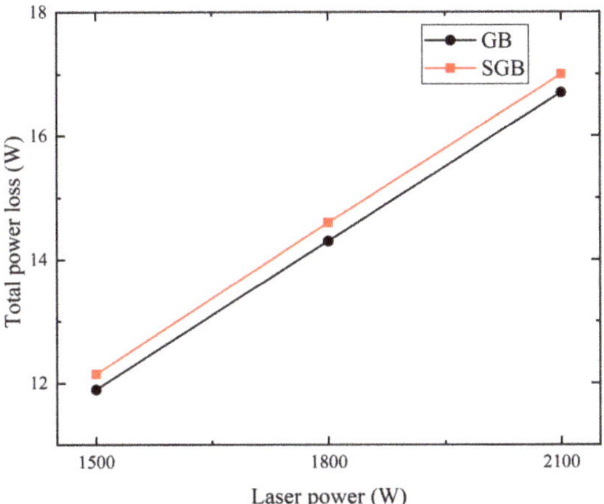

Figure 6. Total power loss under different laser power for the GB and SGB cases (powder feeding rate: 5 g/min; laser spot: 1.6 mm; scanning speed: 10 mm/s).

The temperature of heated powder particles on the deposition surface for the GB and SGB cases are shown in Figure 7a,b, respectively. It is clearly seen that the temperature has a similar distribution with the original laser intensity and it does not exceed the melting point, which indicates that the powder has a cooling effect for the molten pool. Moreover, the peak temperature of powder is located around the laser beam center and the value is 749.4 K for the GB case while it is 581.1 K for the SGB case.

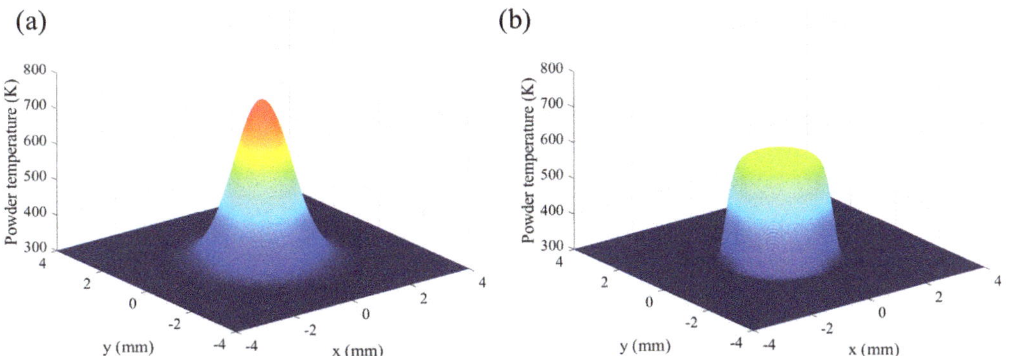

Figure 7. Powder temperature on the liquid/gas interface under (**a**) GB and (**b**) SGB (laser power: 1500 W; powder feeding rate: 5 g/min; laser spot: 1.6 mm; scanning speed: 10 mm/s).

Figure 8 shows the peak temperature of powders under different laser power for the GB and SGB cases. It is found that the peak temperature grows linearly with the increase in the laser power for both the cases. However, the peak temperature is always lower for the SGB case and the rising speed is also slightly lower.

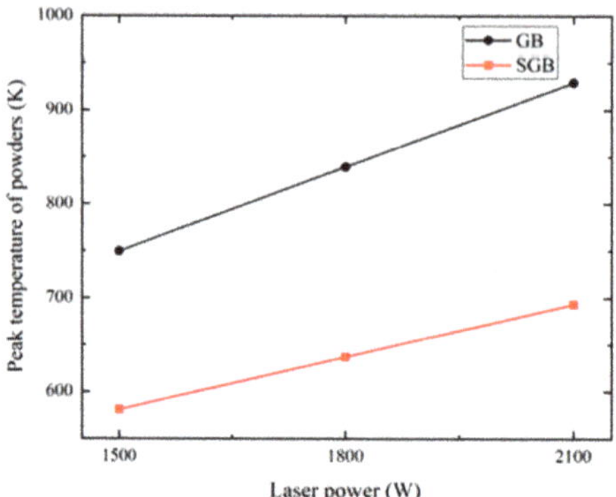

Figure 8. Peak temperature of powders under different laser power for the GB and SGB cases (powder feeding rate: 5 g/min; laser spot: 1.6 mm; scanning speed: 10 mm/s).

4.2. Temperature and Velocity Field

To understand the evolution of temperature field in the L-DED process, the peak temperature of the molten pool under different times has been plotted in Figure 9. The peak temperature sharply rises in the initial stage before it exceeds the solidus temperature. After the molten pool forms, the temperature rises with a slower rate and gradually attains a quasi-steady state. Compared with the GB heat source, the temperature rising rate for the SGB case is slower and there is a 20 ms delay for the molten pool temperature to exceed the solidus temperature. It is found that the peak temperature has an obvious deviation after 200 ms and the calculated maximum temperature for the SGB case is 90 K lower than that for the GB case. It is indicated that the peak temperature of the system can be reduced using the SGB heat source.

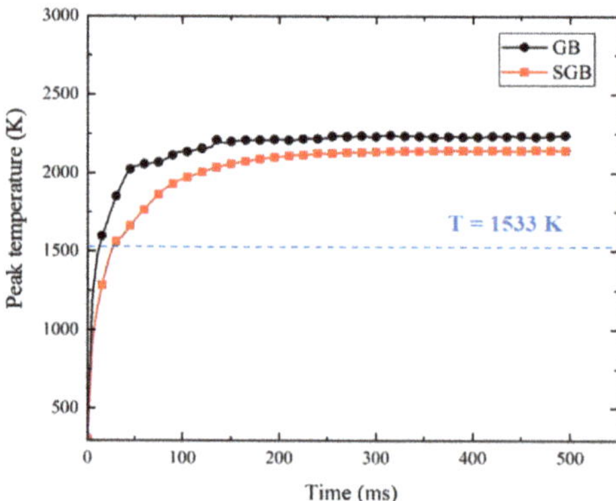

Figure 9. Evolution of maximum temperature in the molten pool.

Figure 10 shows the temperature and velocity field of the molten pool at 500 ms after the molten pool is stabilized. Corresponding laser power (P), scanning speed (v_s) and powder feeding rate (M_p) are 1500 W, 10 mm/s and 5 g/min, respectively. Temperature is indicated by the contour, as shown in the figure, and the arrows represent the fluid flow direction. The black isotherm of 1533 K represents the solid temperature, and three blue isotherms are set at 1700 K, 1900 K and 2100 K, respectively. The denser isotherms indicate a greater change in temperature while sparser isotherms mean that temperature is more uniform. According to Figure 10a,b, the sparse isotherm behind the laser beam center indicates that heat accumulation occurs at the rear side while the dense isotherm at the front side indicates that the temperature is nonuniform in this region. An outward flow is spotted around the free surface of the molten pool, which is caused by the thermal capillary force that drives the liquid metal from low surface tension region to high surface tension region. Moreover, the fluid is more active for the GB case compared with that for the SGB case and the corresponding maximum fluid velocity are 0.32 m/s and 0.28 m/s, respectively.

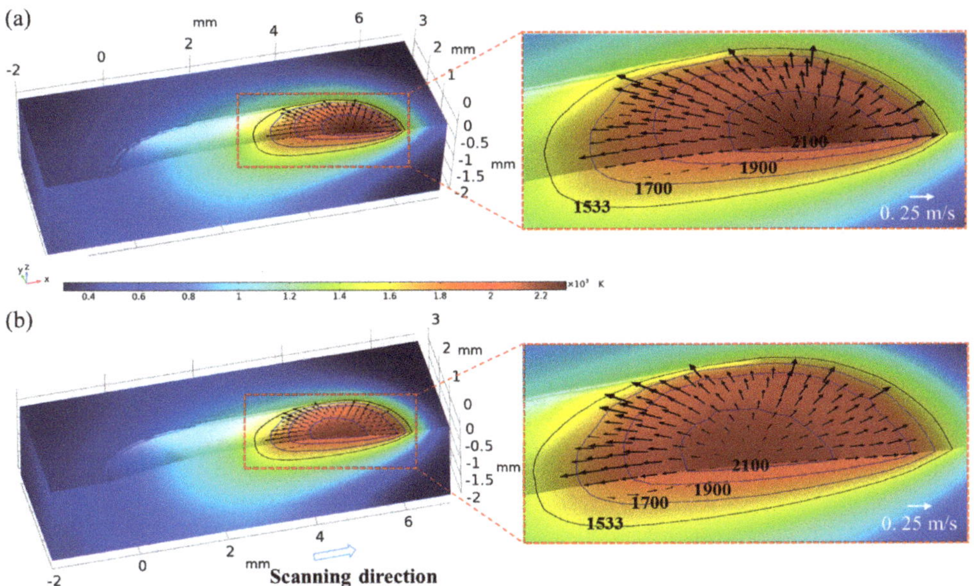

Figure 10. Temperature and velocity field at 500 ms for (**a**) GB and (**b**) SGB.

Figure 11 shows the temperature and velocity field at 500 ms at xz plane (y = 0). Two opposite vortices are found at the two sides of the reference line and the magnitude of the rear one is bigger. The fluid in the center of the vortex is at a standstill which means that no fluid motion occurs in this region. Moreover, the peak magnitude of vortices ($\nabla \times \mathbf{u}$) for the GB case is approximately 1.5 time that for the SGB case.

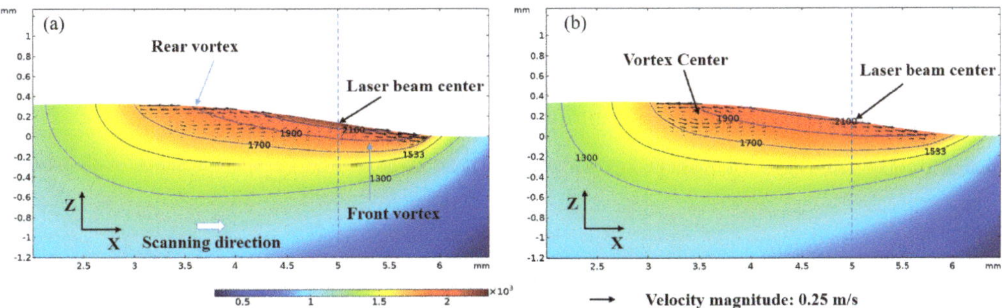

Figure 11. Temperature and velocity field under (**a**) GB and (**b**) SGB at xz plane(y = 0).

In order to analyze the temperature and fluid distribution on the free surface, Figure 12a,b show the local temperature on the deposited surface for line 1 and line 2 at 500 ms, respectively. The deviation of the local temperature in the solid zone (temperature below 1533 K) for the GB and SGB cases is less than 3% behind the laser beam center, while a slight difference is observed in front of the center of laser beam. According to Figure 12b, temperature slightly decreases along y direction around the beam center while it decreases rapidly when it is close to the boundary. The calculated length and width under the SGB condition are 0.17 mm and 0.19 mm wider than the GB case, respectively. Figure 12c,d exhibit the fluid velocity on the free surface for the GB and SGB cases, respectively. The fluid velocity has two peak values at two sides of the laser beam center and the deviation between the rear and front side is less than 5% for the SGB case while it is larger for the GB case. Based on Figure 12d, the fluid velocity starts to increase from the laser beam center and it decreases sharply when it is close to the molten pool boundary.

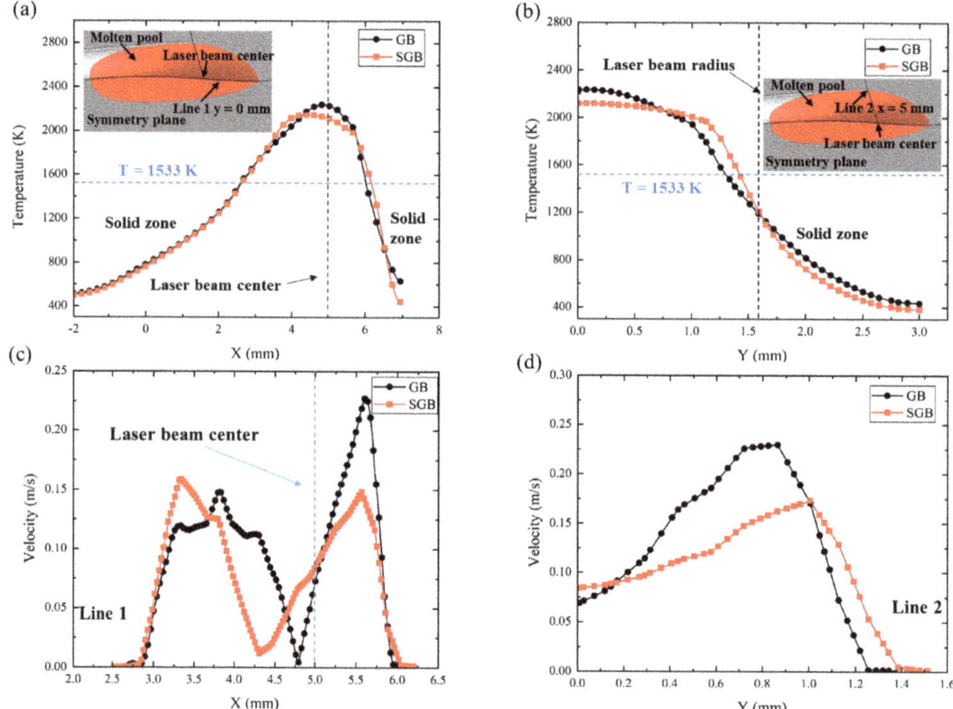

Figure 12. Temperature and velocity distribution along different line at 500 ms (**a**) temperature along line 1, (**b**) temperature along line 2, (**c**) velocity along line 1, (**d**) velocity along line 2.

To estimate the uniformity of temperature distribution of molten pool, the temperature non-uniformity δ_T is calculated. The formula of temperature non-uniformity is as follows [10]:

$$\delta_T = \frac{\sum\limits_i V_i \left| \frac{T_i - T_{ave}}{T_{ave}} \right|}{\sum\limits_i V_i} \tag{31}$$

where, i deontes the ith cell and V_i represents the cell volume. T_{ave} is the average temperature of the molten pool and is calculated using volume averaging method.

It should be noted that the temperature non-uniformity is calculated inside the molten pool whose temperature is above the solid temperature. The greater uniform temperature inside the molten pool leads to the smaller value of δ_T. The calculated temperature non-uniformity is 0.073 and 0.072 for the SGB and GB cases, respectively. Even though the distribution of energy input is quite different, temperature field inside the molten pool for these two cases is almost the same. Additionally, the calculated temperature non-uniformity is used to help explain the fluid motion in next section.

A parametric study has been performed to investigate the influence of the laser power on the evolution of temperature and velocity in the molten pool. Two different cases under the SGB energy input have been analyzed here and the input power is 1800 W, and 2100 W, respectively. Figure 13a,b shows the temperature and velocity field under different laser power after the molten pool is stabilized, respectively. It is found that the increase in molten pool size is due to the increase in heat input. Moreover, a more melting volume above 2300 K is observed as the laser power increase.

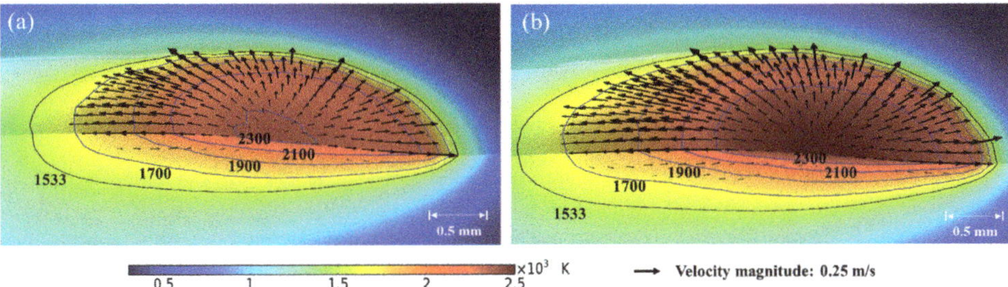

Figure 13. Temperature and velocity field under different laser power for the SGB case at 500 ms (**a**) 1800 W, (**b**) 2100 W.

Based on the simulated results, the melting volume, peak temperature, peak velocity and temperature non-uniformity have been presented in Table 3. It is clearly seen that the melting volume, peak temperature and peak velocity are larger due to the increase in laser power. The increase in fluid velocity can be attributed to less uniform temperature, which is consistent with the calculated temperature non-uniformity.

Table 3. Melting volume, peak temperature, peak velocity and temperature non-uniformity under different laser power.

Case	Melting Volume (m^{30})	Peak Temperature (K)	Peak Velocity (m/s)	Temperature Non-Uniformity (-)
SGB1500	1.32×10^{-9}	2149	0.28	7.29×10^{-2}
SGB1800	2.05×10^{-9}	2322	0.34	8.56×10^{-2}
SGB2100	2.86×10^{-9}	2481	0.39	9.60×10^{-2}

4.3. Dimensionless Analysis

Thermal conduction and convection are two mechanism of heat transfer in the L-DED process. To find which one is dominate, a non-dimensional Peclet number (*Pe*) is used to evaluate the relative importance and it is defined as the ratio of momentum and energy diffusion term. The related formula is as follows:

$$Pe = \frac{\rho C_p \overline{U} L_c}{k} \tag{32}$$

where, $\overline{U}$ is the average fluid velocity within the molten pool, and it is calculated by using volume average method and the formula is expressed as:

$$\overline{U} = \frac{\sum_i V_i u_i}{\sum_i V_i} \tag{33}$$

L_c is the equivalent molten pool radius which is calculated by converting the molten pool volume to the equivalent hemisphere and the formula is given as:

$$r_a = \sqrt[3]{\frac{3V_m}{2\pi}} \tag{34}$$

The average liquid velocity within the molten pool and calculated Peclet number (Pe) have been plotted in Figure 14. The fluid velocity is zero in the initial stage because the material does not reach the melting point and it is kept at a certain level after the system reaches a stabilized condition. Moreover, the average liquid velocity under the GB condition is approximately 16% higher than that for the SGB case, which means that the entire fluid motion is more active for the GB case. Based on Figure 14b, the Pe number has the same tendency with the average velocity. Except for the beginning stage, the value of Pe exceeds the unit, underlining that thermal convection dominated the mechanism of heat transfer. The calculated melting volume for the SGB case is 1.32×10^{-9} m^3, while it is 1.19×10^{-9} m^3 for the GB case, indicating an almost 10% bigger melt volume for the SGB case, which results in a larger apparent molten pool radius. Although the characteristic length under the SGB condition is about 3.4% wider than the GB case, the calculated Pe number under the GB condition is larger (Figure 14b), which is attributed to the larger magnitude of fluid velocity. It is noted that the Pe number for the GB case is approximately 12% larger than that for the SGB case at 500 ms, meaning that the thermal convection has a larger influence on heat transfer compared to thermal conduction.

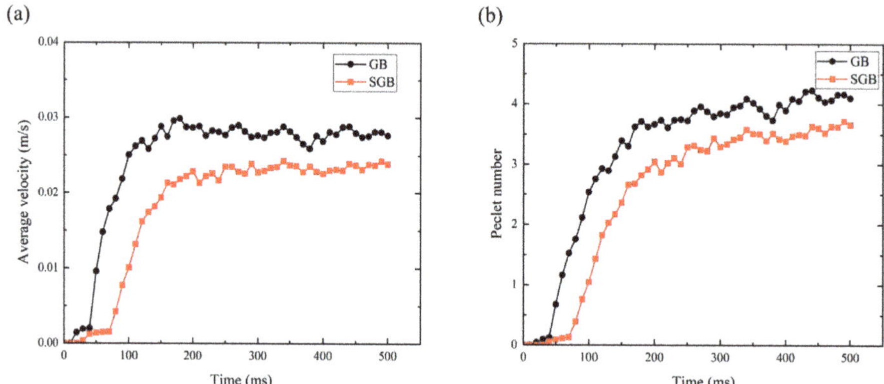

Figure 14. Average liquid velocity (**a**) and the Pe number (**b**) against time for the GB and SGB cases.

Thermal capillary force results in the fluid flow on the free surface is known as Marangoni–Benard convection. Buoyancy results in the fluid flow in the entire molten pool is known as Rayleigh–Benard convection.

Marangoni number (Ma) is defined as the ratio of surface shear stress and viscous shear stress which evaluates the Marangoni shear stress of the metal fluid. It is given as [6]:

$$Ma = \frac{\rho \Delta T_{\max} L_c |\gamma|}{\mu \alpha} \tag{35}$$

where, $\Delta T_{\max}$ represents the deviation between the melting point and peak temperature inside the molten pool. α is the thermal diffusivity.

Rayleigh number (Ra) is used to express the relative strength of buoyancy and viscous force within the molten pool and it is given as [36]:

$$Ra = \frac{\rho \beta g \Delta T_{\max} L_c^3}{\mu \alpha} \tag{36}$$

Figure 15a,b show the variation of Marangoni and Rayleigh number with time within the molten pool under the GB and SGB conditions, respectively. It should be noticed that the Marangoni number is much larger than the unit, meaning that the thermal capillary force is much larger than the viscous force for both cases. From Figure 15a, it is found that the Marangoni number under SGB condition is always lower than that under the GB condition which is caused by the uniform energy. Moreover, thermal capillary for the GB case has more significant influence on the molten pool dynamics, which is caused by the non-uniform heat input on the deposited surface. From Figure 15, it is found that the Ra number is approximately three orders of magnitude smaller than the Ma number, which is mainly due to the small characteristic length (of the order of 10^{-3} m) of the molten pool. In this respect, Marangoni effect dominates the fluid flow pattern in the L-DED process.

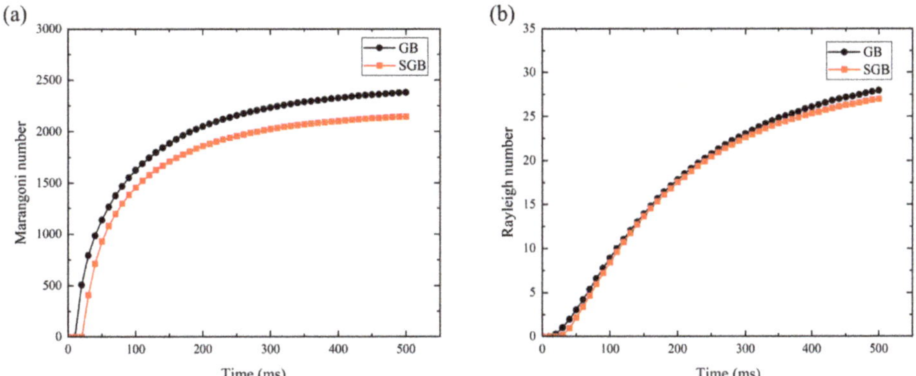

Figure 15. Plot of Ma number (**a**) and Ra number (**b**) against time for the GB and SGB cases.

It is worth mentioning that the Prandtl number (Pr = $\mu c_p/k$), which is closely related to the location of vortices, has a significant influence on the fluid dynamics and further on the appearance of the molten pool [37]. Pr number is used to describe the boundary layer during the molten pool evolution. The calculated Pr number (in this case Pr = 0.14) is less than unit, which indicates that the momentum dissipation is the primary mechanism.

The Fourier number (*Fo*) is a dimensionless number which is related to the two opposite mechanisms of heat dissipation and heat accumulation. A high Fourier number indicates that the heat within the molten pool can be dissipated more easily and the formula can be expressed as [6]:

$$Fo = \frac{\alpha}{v_s L_c} \tag{37}$$

For the GB and SGB cases, the calculated Fourier numbers are 0.544 and 0.526 at 500 ms, respectively. The former is approximately 3.4% larger than the latter, which is attributed to the bigger melting volume. Thus, less Fo number indicates that the molten pool for the SGB case has a weaker heat dissipation capability, and on the other hand, more heat is accumulated inside the molten pool rather than dissipated to the no-melting region. To explain this process, the thermal transfer towards the no-melt region has been calculated on three selected planes, as shown in Figure 16. Here, the net output power (*NOP*) is defined as the integration of heat flux through the selective plane, the formula can be expressed as:

$$NOP = \sum_i A_i q_i \tag{38}$$

where, A_i represents the area of *i*th cell in the selective plane and q_i denotes the heat flux through the selective plane. Based on Figure 16, although the predicted NOP has the same tendency in the two cases at three planes, an obvious difference can be discovered at plane A and B while its difference is less than 3% at plane C. Hence, the heat dissipation capability is weaker for the SGB case due to its relatively lower fluid velocity, as shown in Figures 12 and 14. It is also found that fluid flow will significantly influence the heat transport between the molten pool and the no-melt region in LPBF [38].

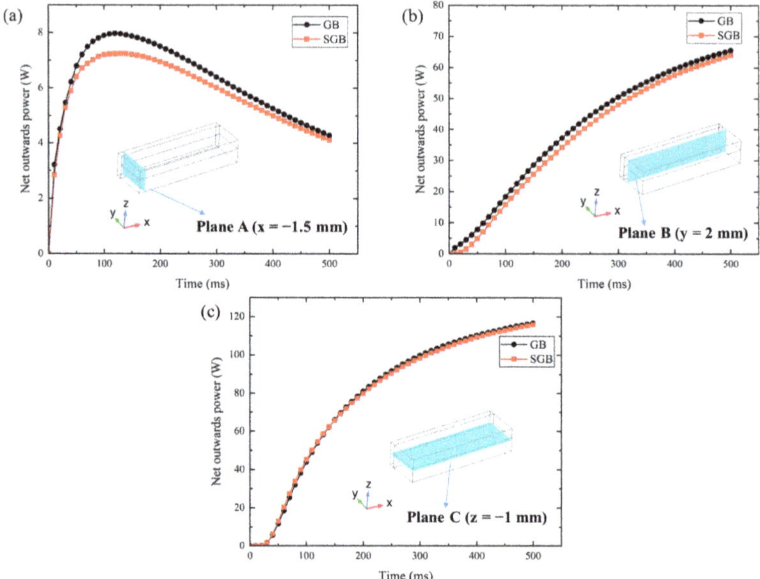

Figure 16. Net output power calculated on a plane versus time (**a**) x = −1.5 mm, (**b**) y = 2 mm, (**c**) z = −1 mm.

For further analysis of the thermal-fluid transport inside the molten pool, the relevant non-dimensional numbers have been calculated and presented in Table 4. It can be found that Pe number is directly related to the heat input and an increase in laser power will strength heat convection. Moreover, an increase in Ma and Ra number indicates that both Marangoni and buoyance effect are strengthened by higher laser power. In addition, because the ratio of Ma number and Ra number is on the order of 100, Marangoni effect is the main physical mechanism dominating the driving forces for liquid flow. From Table 4, it is evident that the Fo number decreases as the heat input increases, which indicates that the heat dissipation capability is weaker when the laser power rises.

Table 4. Dimensionless numbers under different laser power.

Case	$\overline{U}$ (cm/s)	Pe	Ra	Ma	Fo
SGB1500	2.4	31.8	26.9	2146.8	0.526
SGB1800	3.0	46.2	53.6	3184.9	0.454
SGB2100	3.4	58.8	89.8	4271.8	0.407

4.4. Solidification Characteristics

Grain morphology of metals during the L-DED process depends on solidification parameters. Temperature gradient (*G*) is a solidification parameter and its direction is normal to the solidification front. The model in this paper is based on the Cartesian coordinate system, thus *G* is expressed as:

$$G = \left(\frac{\partial T}{\partial x}, \frac{\partial T}{\partial y}, \frac{\partial T}{\partial z}\right) \tag{39}$$

Growth velocity R is another solidification parameter which is normal to the liquid/gas interface and its magnitude can be expressed as:

$$R = v_s \cdot \frac{\partial T}{\partial x} / G \tag{40}$$

The solidification parameter $G \times R$ is related to the grain size, representing the instantaneous cooling rate on the solidification front. G/R is another solidification parameter which has important relevance to the microstructural morphology [39]. G and R along line 1 for the GB and SGB cases are shown in Figure 17a,b, respectively. It is found that G first decreases then a slight increase is observed around the top surface for both the cases. The maximal magnitude of G is located at the bottom region and the corresponding value is 8.1×10^5 K/m and 8.4×10^5 K/m, respectively. The slight increase in G at the top region is mainly due to the thermal exchange, including convection and radiation with the external environment. Growth velocity of the solidification front increases from 0.51 mm/s and 0.50 mm/s at the bottom region to 9.8 mm/s and 10 mm/s at the top region for the GB and SGB cases, respectively. The top region in the center has a growth velocity which is almost the same with the scanning speed, meaning that the local growth direction is consistent with the laser moving direction. However, R at the bottom region is approximately 3% of that at the top for both cases. G and R along line 2 are plotted in Figure 17c,d, respectively. The variation tendency of G and R along line 2 is similar to those along line 1. However, temperature gradient along line 2 is always higher for the SGB case, and the growth velocity in the central region for the SGB case is larger than that for the GB case, while it is almost the same on two sides.

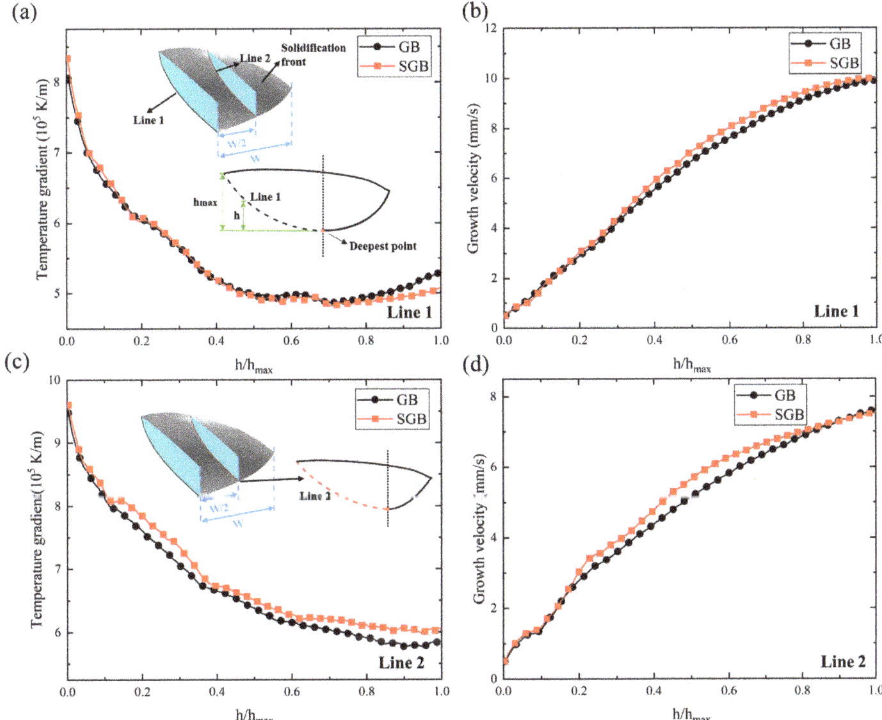

Figure 17. Temperature gradient and growth velocity under the laser power of 1500 W for the GB and SGB cases along different path: (**a**) temperature gradient for line 1, (**b**) growth velocity for line 1, (**c**) temperature gradient for line 2 and (**d**) growth velocity for line 2.

Figure 18a,b show the change in $G \times R$ at the line 1 and line 2, respectively, with the relative height of the solidification front and the corresponding laser power is 1500 W. $G \times R$ increases from 412 K/s to 5264 K/s from the bottom region to the top region for the GB case while it increases from 420 K/s to 5069 K/s for the SGB case. From Figure 18a, $G \times R$ is lower at the top region for the SGB case, while it is larger in the middle region. However, the relative difference is less than 5%. According to Figure 18b, $G \times R$ increases with the increase in relative height and it reaches 4455 K/s and 4561 K/s for the GB and SGB cases, respectively. Moreover, it can be found that $G \times R$ along line 2 is higher for the SGB case, which indicates that the grain size could be smaller for the SGB case. The higher magnitude of $G \times R$ is attributed to the higher temperature gradient and growth velocity according to Figure 17c,d.

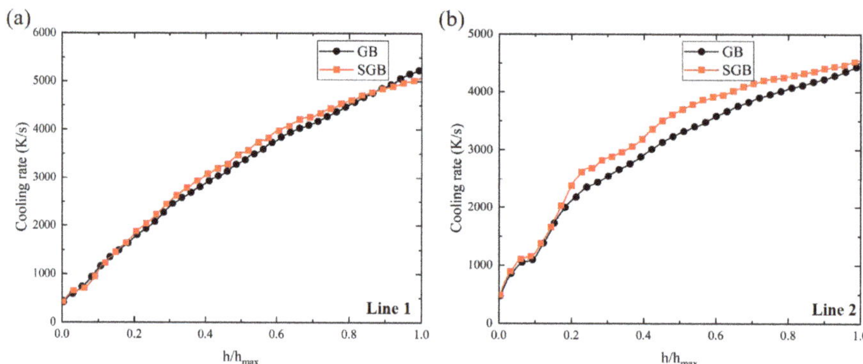

Figure 18. Cooling rate under the laser power of 1500 W for GB and SGB cases along different path: (**a**) line 1 and (**b**) line 2.

Figure 19a,b show the morphology factor (G/R) along the line 1 and line 2, respectively. It is found that with the increase in relative height, the morphology factor rapidly first decreases, after that, it decreases with a slower speed until reaching the molten pool surface. Thus, it can be predicted that columnar growth appears more likely at the bottom region for both the cases. The relative error in this region for the two lines is less than 5%. Comparing the morphology factor along the different paths, it is found that G/R is larger along line 2, which is mainly caused by the higher magnitude of G and R according to Figure 17. Based on Figure 19a, the maximum value of the solidification parameter G/R is located at the bottom region and the corresponding values are 1.59×10^9 K·s/m^2 and 1.68×10^9 K·s/m^2 for the GB and SGB cases, respectively.

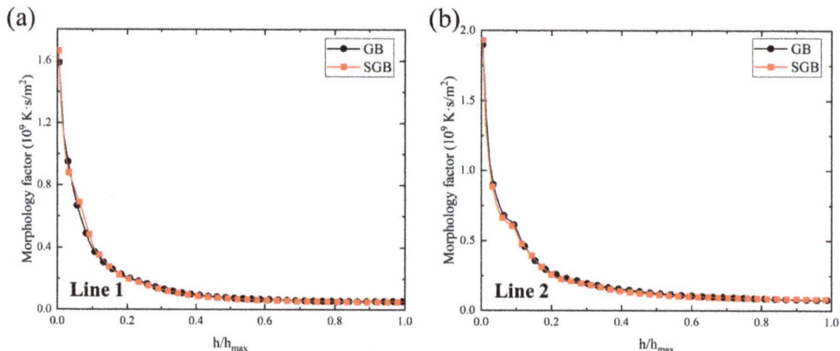

Figure 19. (**a**) Morphology factor under the laser power of 1500 W for GB and SGB cases along different path: (**a**) line 1 and (**b**) line 2.

Figure 20 shows the *G* and *R* along line 1 plotted on a reference solidification map, indicating the influence of laser power on grain size and morphology [40]. The arrows indicating "finer grain" correspond to the direction of increasing cooling rate, which indicates that the grains can be refined by lowering the heat input. It is evident that grain morphology along line 1 under different laser power is columnar especially for the bottom region, which is mainly caused by the smaller *R* and higher *G*.

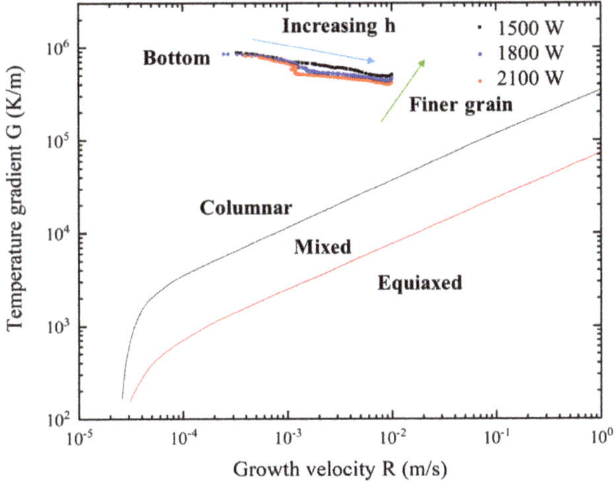

Figure 20. Predicated temperature gradient and growth velocity along line 1 plotted on reference solidification map [40] under different laser power.

To further analyze the solidification parameters under different cases, the average solidification parameters are calculated using the area-weighted average method and its formula is as follows:

$$\overline{f} = \frac{\sum\limits_{i} A_i f_i}{\sum\limits_{i} A_i} \tag{41}$$

where, A_i represents the area of *i*th cell of the solidification front and f_i denotes the solidification parameter. In this equation, f_i should be replaced by G, R, $G \times R$ and G/R.

The calculated average solidification parameters are presented in Table 5. It is found that the average cooling rate is 11.1% higher for the SGB case under the laser power of 1500 W, which means that the grains could be finer for the SGB case in the L-DED process. Moreover, as the laser power increases from 1500 W to 2100 W, the average cooling rate decreases from 2.40×10^3 K/s to 1.71×10^3 K/s, which indicates that the average grains size in the clad layer would become larger with the increase in the energy input.

Table 5. Calculated average of the solidification parameters.

Case	Temperature Gradient (K/m)	Growth Velocity (m/s)	Cooling Rate (K/s)	Morphology Factor (K·s/m^2)
GB1500	7.72×10^5	3.16×10^{-3}	2.16×10^3	6.16×10^8
SGB1500	7.98×10^5	3.36×10^{-3}	2.40×10^3	6.60×10^8
SGB1800	7.76×10^5	2.93×10^{-3}	1.99×10^3	1.66×10^9
SGB2100	7.38×10^5	2.69×10^{-3}	1.71×10^3	1.72×10^9

Inconel 718 is a typical alloy that has large solidification ranges due to its several alloying elements [41]. During the solidification process, the undercooled region is formed and several crystal structures, such as stable planar crystals, unstable cellular crystals and dendrites, may be produced in this region. [39]. Research by previous scholars [15,16] have revealed that the solidification morphology of Inconel 718 during rapid solidification processes might be either cellular or dendritic, which depends on the experimental parameters. The transformation of columnar to equiaxed transition (CET) is dependent on the constitutional supercooling (CS) [42], which is usually described by the solidification parameter G/R. Lower magnitude of G/R results in more CS in the solidification front, which brings more possibility for the equiaxed grain growth. As the undercooling in the solidification front exceeds the required subcooling of the nucleation, equiaxed crystals can precede columnar nucleate.

In considering multi-component solute systems of Inconel 718 alloy, the constitutional tip undercooling is related to the solute concentration [43].

$$\Delta T_c = \sum_{i=1}^{n} m_i (C_{0,i} - C_{l,i}^*) \tag{42}$$

where, m_i is the slope of the liquidus. Additionally, $C_{0,i}$ is the nominal concentration of the alloys and $C_{l,i}^*$ denotes the concentration at the tip in the liquid. The thermodynamics and kinetics of the CET are simplified by using an empirical relation, as presented by Gäumann et al. [43]:

$$\frac{G^n}{R} = a \left[\sqrt[3]{\frac{-4\pi N_0}{3\ln(1-\phi)} \frac{1}{n+1}} \right]^n \tag{43}$$

where, a and b are constants related to the alloy, N_0 denotes the nucleation density, and ϕ is the volume fraction of equiaxed grains which is also called stray grains (SGs). According to Hunt's research [42], the system can be considered fully columnar growth when the volume fraction of the SGs is below 0.66%. While the volume fraction exceeds 49%, the system is considered as fully equiaxed growth. Between these two regions, there is mixed columnar/equiaxed grain growth. In this respect, the pattern of grain growth can be predicted by calculating G and R.

The fraction of the SGs can be calculated by rearranging Equation (43).

$$\phi = 1 - \exp\left\{ \frac{-4\pi N_0}{3} \left(\frac{1}{(n+1)(G^n/aR)} \right)^3 \right\} \tag{44}$$

It is observed that the fraction of the SGs depends on the material and solidification parameters. The constants associated with alloys and nucleation density are using the parameters reported by Gäumann et al. [43] for a similar nickel-base superalloy while the solidification parameters are calculated in the 3D numerical model. The average volume fraction of the SGs is calculated using the same method as the average solidification parameters and it is given as [44]:

$$\overline{\phi} = \frac{\sum\limits_{i} A_i \phi_i}{\sum\limits_{i} A_i} \tag{45}$$

Three predicted volume fraction maps of the SGs under different laser power are presented in Figure 21. It is shown the formation possibility of the SGs and the value of ϕ in the central-top region is the maximum for three different cases. Moreover, the predicted maximal volume fraction of the SGs reaches 0.3 for the SGB1500 case, while it reaches 0.47 the SGB2100 case. Thus, it can be predicted that ϕ could increase as the heat input increases.

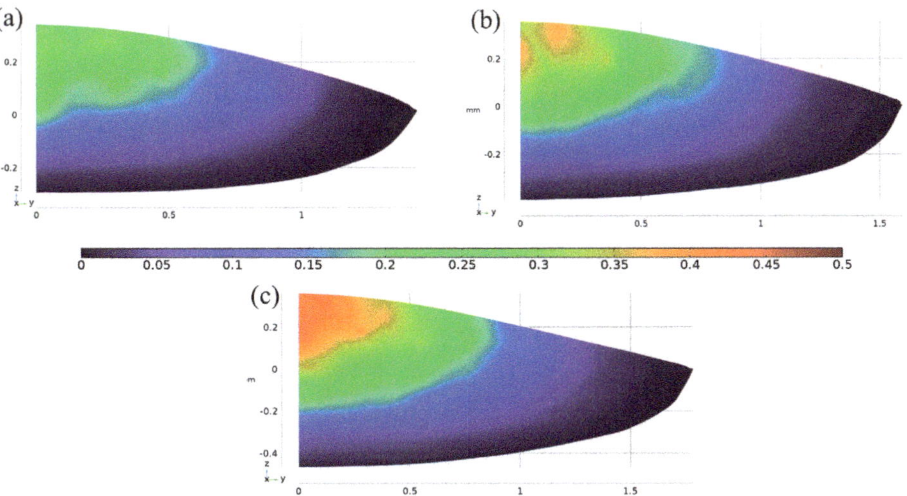

Figure 21. Predicted volume fraction of the SGs on the solidification plane for the SGB case under different laser powers (**a**) 1500 W (**b**) 1800 W (**c**) 2100 W.

The calculated average volume fraction of the SGs under different laser powers is plotted in Figure 22. It is found that the predicted average fraction of the SGs is 5.6% with a laser power of 1500 W, while it increases to 6.6% with a laser power of 2100 W. Thus, low heat input conditions are helpful to avoid the formation of the SGs and promote directional grain growth.

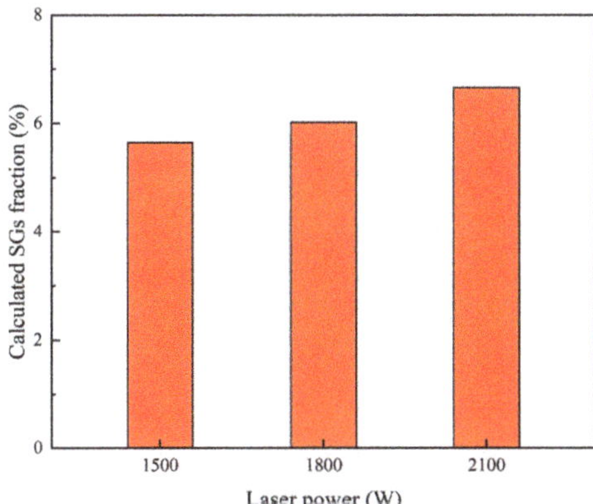

Figure 22. Predicted average volume fraction of the SGs under different laser powers.

4.5. Validation of the Model

The cross-sectional morphology of the clad layer, derived from experiments and computation, is plotted on Figure 23. The results show that predicted geometry by numerical simulation matches well with the experiment results.

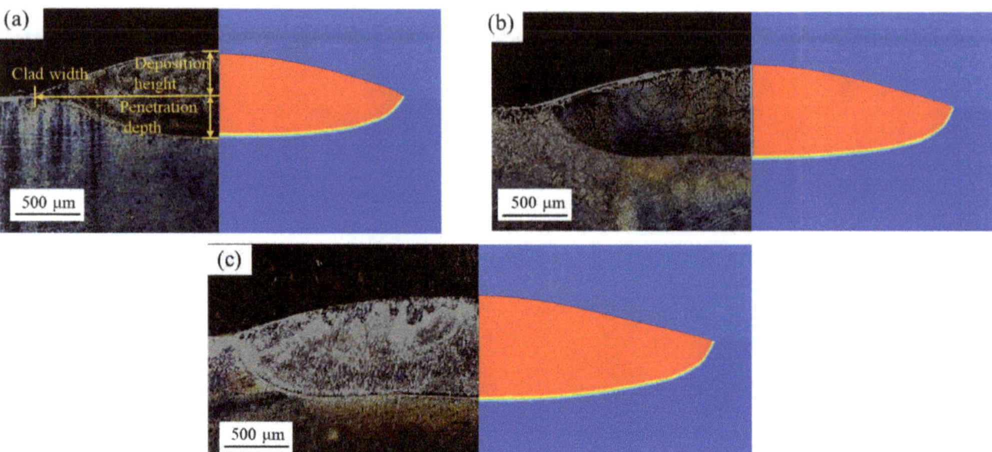

Figure 23. Experimental and computed cross-sections of the clad layer for the SGB case at different laser powers: (**a**) 1500 W, (**b**) 1800 W and (**c**) 2100 W.

Table 6 shows the comparison of geometric parameters for cross-section of the clad layer between the experimental and simulated results. It is found that the width and depth of the cladding layer increase as the heat input increases. However, the deposition height is almost the same for three cases. This phenomenon is due to two opposite factors, the flow of metal liquid and the "catch" efficiency of the liquid metal. On one hand, increasing the heat input leads to a larger molten pool size which could catch more powder particles, thus the deposition height is larger. On the other hand, the flow of metal liquid is also strengthened with the increase in heat input, causing liquid metal from the center to flow towards the edge.

Table 6. Experimental and computed geometric parameters.

Case	Clad Width (mm)			Penetration Depth (mm)			Deposition Height (mm)		
	Exp.	Sim.	Error (%)	Exp.	Sim.	Error (%)	Exp.	Sim.	Error (%)
SGB1500	1.46	1.45	0.6%	0.32	0.30	6.3%	0.35	0.34	2.3%
SGB1800	1.57	1.62	3.2%	0.39	0.40	1.8%	0.34	0.36	4.7%
SGB2100	1.88	1.78	5.4%	0.46	0.47	2.0%	0.34	0.36	7.0%

The results show that the prediction error of the clad width is less than 5.4%, penetration depth error is less than 6.3% and deposition height error is less than 7.0%, which means that the prediction of the numerical simulation is good. From the validations, it is observed that the computed results have some differences with the experimental results, which can be attributed to many factors, such as neglect of shielding gas and vaporization.

5. Conclusions

The thermal-fluid transfer behavior and solidification characteristics under two types of laser beam, the GB and SGB, are analyzed utilizing an improved three-dimensional numerical model. Based on the numerical simulation, the following specific conclusions can be drawn:

1. The total attenuation of laser intensity has a Gaussian-like shape for both cases. During the laser–powder interaction, the temperature of the heated powder particles is lower than the melting point of the metal. Moreover, the particle temperature on the liquid/gas interface has a similar distribution with its original laser intensity.

2. The laser beam profile has a significant influence on the temperature field and fluid flow in the molten pool. Compared with the GB case, the peak temperature for the SGB case is approximately 90 K lower, and the average liquid velocity within the molten pool is about 14% lower.

3. The main heat dissipation mechanism in the molten pool is convection rather than conduction, because Pe number is much larger than the unit. The Peclet number for the GB case is approximately 12% larger than that for the SGB case, meaning that the fluid flow has a greater effect on heat transfer compared with thermal conduction. Moreover, thermal capillary force is the main driving force for the GB and SGB cases.

4. The average temperature gradient and cooling rate are both higher for the SGB case which indicates that the grains could be finer for the SGB case in the L-DED process. Additionally, the higher magnitude of morphology factor for the SGB case means that it has more possibility for columnar growth.

Author Contributions: Conceptualization, B.C., X.H., G.Y., Z.L., S.L. and C.T.; methodology, B.C., X.H., Y.B., B.D. and S.L.; software, B.C., X.H. and Z.L.; validation, G.Y., X.H. and Z.L.; formal analysis, B.C. and C.T.; investigation, B.C., B.D. and X.H.; resources, G.Y., X.H., Y.B. and S.L.; data curation, B.C., Y.B. and B.D.; writing—original draft preparation, B.C.; writing—review and editing, X.H.; visualization, B.C., X.H. and C.T.; supervision, X.H. and G.Y.; project administration, X.H. and G.Y.; funding acquisition, Z.L. and S.L. All authors have read and agreed to the published version of the manuscript.

Funding: This work was supported by National Natural Science Foundation of China (No. 12202448), High-level Innovation Research Institute Program of Guangdong Province (No. 2020 B0909010003) and Research Project of Guangdong Aerospace Research Academy (GARA2022001000).

Institutional Review Board Statement: Not applicable.

Informed Consent Statement: Not applicable.

Data Availability Statement: Not applicable.

Conflicts of Interest: The authors declare no conflict of interest.

References

1. Gunen, A.; Gurol, U.; Kocak, M.; Cam, G. A new approach to improve some properties of wire arc additively manufactured stainless steel components: Simultaneous homogenization and boriding. *Surf. Coat Tech.* **2023**, *460*, 129395. [CrossRef]

2. Shao, J.; Yu, G.; Li, S.; He, X.; Tian, C.; Dong, B. Crystal growth control of Ni-based alloys by modulation of the melt pool morphology in DED. *J. Alloys Compd.* **2022**, *898*, 162976. [CrossRef]

3. Lin, S.; Chen, K.; He, W.; Tamura, N.; Ma, E. Custom-designed heat treatment simultaneously resolves multiple challenges facing 3D-printed single-crystal superalloys. *Mater. Des.* **2022**, *222*, 111075. [CrossRef]

4. Çam, G. Prospects of producing aluminum parts by wire arc additive manufacturing (WAAM). *Mater. Today Proc.* **2022**, *62*, 77–85. [CrossRef]

5. Kruth, J.-P.; Leu, M.C.; Nakagawa, T. Progress in Additive Manufacturing and Rapid Prototyping. *Cirp Ann.* **1998**, *47*, 525–540. [CrossRef]

6. DebRoy, T.; Wei, H.L.; Zuback, J.S.; Mukherjee, T.; Elmer, J.W.; Milewski, J.O.; Beese, A.M.; Wilson-Heid, A.; De, A.; Zhang, W. Additive manufacturing of metallic components—Process, structure and properties. *Prog. Mater. Sci.* **2018**, *92*, 112–224. [CrossRef]

7. Zhang, X.; Li, W.; Liou, F. Damage detection and reconstruction algorithm in repairing compressor blade by direct metal deposition. *Int. J. Adv. Manuf. Technol.* **2018**, *95*, 2393–2404. [CrossRef]

8. Lee, J.-Y.; Nagalingam, A.P.; Yeo, S.H. A review on the state-of-the-art of surface finishing processes and related ISO/ASTM standards for metal additive manufactured components. *Virtual Phys. Prototyp.* **2021**, *16*, 68–96. [CrossRef]

9. Wen, S.; Shin, Y.C. Modeling of transport phenomena during the coaxial laser direct deposition process. *J. Appl. Phys.* **2010**, *108*, 044908. [CrossRef]

10. Dortkasli, K.; Isik, M.; Demir, E. A thermal finite element model with efficient computation of surface heat fluxes for directed-energy deposition process and application to laser metal deposition of IN718. *J. Manuf. Process* **2022**, *79*, 369–382. [CrossRef]

11. Sun, Z.; Guo, W.; Li, L. Numerical modelling of heat transfer, mass transport and microstructure formation in a high deposition rate laser directed energy deposition process. *Addit. Manuf.* **2020**, *33*, 101175. [CrossRef]

12. Gan, Z.; Yu, G.; He, X.; Li, S. Numerical simulation of thermal behavior and multicomponent mass transfer in direct laser deposition of Co-base alloy on steel. *Int. J. Heat Mass Transf.* **2017**, *104*, 28–38. [CrossRef]

13. Li, W.; Kishore, M.N.; Zhang, R.Y.; Bian, N.; Lu, H.B.; Li, Y.Y.; Qian, D.; Zhang, X.C. Comprehensive studies of SS316L/IN718 functionally gradient material fabricated with directed energy deposition: Multi-physics & multi-materials modelling and experimental validation. *Addit. Manuf.* **2023**, *61*, 103358.

14. Bayat, M.; Nadimpalli, V.K.; Biondani, F.G.; Jafarzadeh, S.; Thorborg, J.; Tiedje, N.S.; Bissacco, G.; Pedersen, D.B.; Hattel, J.H. On the role of the powder stream on the heat and fluid flow conditions during Directed Energy Deposition of maraging steel—Multiphysics modeling and experimental validation. *Addit. Manuf.* **2021**, *43*, 102021. [CrossRef]

15. Song, J.; Chew, Y.; Bi, G.; Yao, X.; Zhang, B.; Bai, J.; Moon, S.K. Numerical and experimental study of laser aided additive manufacturing for melt-pool profile and grain orientation analysis. *Mater. Des.* **2018**, *137*, 286–297. [CrossRef]

16. Wang, C.; Zhou, J.; Zhang, T.; Meng, X.; Li, P.; Huang, S. Numerical simulation and solidification characteristics for laser cladding of Inconel 718. *Opt. Laser Technol.* **2022**, *149*, 107843. [CrossRef]

17. Wu, J.; Zheng, X.; Zhang, Y.; Ren, S.; Yin, C.; Cao, Y.; Zhang, D. Modeling of whole-phase heat transport in laser-based directed energy deposition with multichannel coaxial powder feeding. *Addit. Manuf.* **2022**, *59*, 103161. [CrossRef]

18. Kaplan, A.F.H. Influence of the beam profile formulation when modeling fiber-guided laser welding. *J. Laser Appl.* **2011**, *23*, 042005. [CrossRef]

19. Han, L.; Liou, F.W. Numerical investigation of the influence of laser beam mode on melt pool. *Int. J. Heat Mass Transf.* **2004**, *47*, 4385–4402. [CrossRef]

20. Ayoola, W.; Suder, W.; Williams, S. Effect of beam shape and spatial energy distribution on weld bead geometry in conduction welding. *Opt. Laser Technol.* **2019**, *117*, 280–287. [CrossRef]

21. Huang, S.; Narayan, R.L.; Tan, J.H.K.; Sing, S.L.; Yeong, W.Y. Resolving the porosity-unmelted inclusion dilemma during in-situ alloying of Ti34Nb via laser powder bed fusion. *Acta Mater.* **2020**, *204*, 116522. [CrossRef]

22. Yuan, W.; Chen, H.; Li, S.; Heng, Y.; Yin, S.; Wei, Q. Understanding of adopting flat-top laser in laser powder bed fusion processed Inconel 718 alloy: Simulation of single-track scanning and experiment. *J. Mater. Res. Technol.* **2022**, *16*, 1388–1401. [CrossRef]

23. Wu, J.; Ren, S.; Zhang, Y.; Cao, Y.; Zhang, D.; Yin, C. Influence of spatial laser beam profiles on thermal-fluid transport during laser-based directed energy deposition. *Virtual Phys. Prototyp.* **2021**, *16*, 444–459. [CrossRef]

24. Bian, Y.; He, X.; Yu, G.; Li, S.; Tian, C.; Li, Z.; Zhang, Y.; Liu, J. Powder-flow behavior and process mechanism in laser directed energy deposition based on determined restitution coefficient from inverse modeling. *Powder Technol.* **2022**, *402*, 117355. [CrossRef]

25. Shao, J.Y.; Yu, G.; He, X.L.; Li, S.X.; Chen, R.; Zhao, Y. Grain size evolution under different cooling rate in laser additive manufacturing of superalloy. *Opt. Laser Technol.* **2019**, *119*, 105662. [CrossRef]

26. Zhang, H.-O.; Kong, F.-R.; Wang, G.-L.; Zeng, L.-F. Numerical simulation of multiphase transient field during plasma deposition manufacturing. *J. Appl. Phys.* **2006**, *100*, 123522. [CrossRef]

27. Spiegel, E.A.; Veronis, G. On the Boussinesq Approximation for a Compressible Fluid. *Astrophys. J.* **1960**, *131*, 442. [CrossRef]

28. Brent, A.D.; Voller, V.R.; Reid, K.J. Enthalpy-Porosity Technique for Modeling Convection-Diffusion Phase-Change—Application to the Melting of a Pure Metal. *Numer. Heat Transf.* **1988**, *13*, 297–318. [CrossRef]

29. Shealy, D.L.; Hoffnagle, J.A. Laser beam shaping profiles and propagation. *Appl. Opt.* **2006**, *45*, 5118–5131. [CrossRef]

30. Liu, Z.; Qi, H. Numerical Simulation of Transport Phenomena for a Double-Layer Laser Powder Deposition of Single-Crystal Superalloy. *Met. Mater. Trans. A* **2014**, *45*, 1903–1915. [CrossRef]

31. Frenk, A.; Vandyoussefi, M.; Wagnière, J.-D.; Kurz, W.; Zryd, A. Analysis of the laser-cladding process for stellite on steel. *Met. Mater. Trans. B* **1997**, *28*, 501–508. [CrossRef]

32. Zhang, T.; Li, H.; Liu, S.; Shen, S.N.; Xie, H.M.; Shi, W.X.; Zhang, G.Q.; Shen, B.N.; Chen, L.W.; Xiao, B.; et al. Evolution of molten pool during selective laser melting of Ti-6Al-4V. *J. Phys. D Appl. Phys.* **2019**, *52*, 55302. [CrossRef]

33. Li, S.; Xiao, H.; Liu, K.; Xiao, W.; Li, Y.; Han, X.; Mazumder, J.; Song, L. Melt-pool motion, temperature variation and dendritic morphology of Inconel 718 during pulsed- and continuous-wave laser additive manufacturing: A comparative study. *Mater. Des.* **2017**, *119*, 351–360. [CrossRef]

34. Wirth, F.; Wegener, K. A physical modeling and predictive simulation of the laser cladding process. *Addit. Manuf.* **2018**, *22*, 307–319. [CrossRef]

35. Khandkar, M.Z.H.; Khan, J.A.; Reynolds, A.P. A Thermal Model of the Friction Stir Welding Process. *Int. Mech. Eng. Congr. Expos.* **2002**, *5*, 115–124.

36. Kumar, A.; Roy, S. Effect of three-dimensional melt pool convection on process characteristics during laser cladding. *Comput. Mater. Sci.* **2009**, *46*, 495–506. [CrossRef]

37. Wei, P.S.; Yeh, J.S.; Ting, C.N.; DebRoy, T.; Chung, F.K.; Lin, C.L. The effects of Prandtl number on wavy weld boundary. *Int. J. Heat Mass Transf.* **2009**, *52*, 3790–3798. [CrossRef]

38. Bayat, M.; Mohanty, S.; Hattel, J.H. A systematic investigation of the effects of process parameters on heat and fluid flow and metallurgical conditions during laser-based powder bed fusion of Ti6Al4V alloy. *Int. J. Heat Mass Transf.* **2019**, *139*, 213–230. [CrossRef]

39. Kou, S. *Welding Metallurgy, Second Edition*; Wiley: Hoboken, NJ, USA, 2003.

40. Nastac, L.; Valencia, J.J.; Tims, M.L.; Dax, F.R. Advances in the solidification of IN718 and RS5 alloys. In Proceedings of the Superalloys 718, 625, 706 and Various Derivatives, Pittsburgh, PA, USA, 17–20 June 2001; pp. 103–112.

41. Knapp, G.; Raghavan, N.; Plotkowski, A.; DebRoy, T. Experiments and simulations on solidification microstructure for Inconel 718 in powder bed fusion electron beam additive manufacturing. *Addit. Manuf.* **2019**, *25*, 511–521. [CrossRef]

42. Hunt, J. Steady state columnar and equiaxed growth of dendrites and eutectic. *Mater. Sci. Eng.* **1984**, *65*, 75–83. [CrossRef]
43. Gäumann, M.; Bezençon, C.; Canalis, P.; Kurz, W. Single-crystal laser deposition of superalloys: Processing–microstructure maps. *Acta Mater.* **2001**, *49*, 1051–1062. [CrossRef]
44. Anderson, T.; DuPont, J.; DebRoy, T. Origin of stray grain formation in single-crystal superalloy weld pools from heat transfer and fluid flow modeling. *Acta Mater.* **2010**, *58*, 1441–1454. [CrossRef]

Article

Prediction of Primary Dendrite Arm Spacing of the Inconel 718 Deposition Layer by Laser Cladding Based on a Multi-Scale Simulation

Zhibo Jin, Xiangwei Kong *, Liang Ma, Jun Dong and Xiaoting Li

School of Mechanical Engineering and Automation, Northeastern University, Shenyang 110819, China
* Correspondence: xwkong@me.neu.edu.cn

Abstract: Primary dendrite arm spacing (PDAS) is a crucial microstructural feature in nickel-based superalloys produced by laser cladding. In order to investigate the effects of process parameters on PDAS, a multi-scale model that integrates a 3D transient heat and mass transfer model with a quantitative phase-field model was proposed to simulate the dendritic growth behavior in the molten pool for laser cladding Inconel 718. The values of temperature gradient (G) and solidification rate (R) at the S/L interface of the molten pool under different process conditions were obtained by multi-scale simulation and used as input for the quantitative phase field model. The influence of process parameters on microstructure morphology in the deposition layer was analyzed. The result shows that the dendrite morphology is in good agreement with the experimental result under varying laser power (P) and scanning velocity (V). PDAS was found to be more sensitive to changes in laser scanning velocity, and as the scanning velocity decreased from 12 mm/s to 4 mm/s, the PDAS increased by 197% when the laser power was 1500 W. Furthermore, smaller PDAS can be achieved by combining higher scanning velocity with lower laser power.

Keywords: multi-scale simulation; primary dendrite arm spacing; phase field method; laser cladding; temperature gradient; nickel-based superalloy

Citation: Jin, Z.; Kong, X.; Ma, L.; Dong, J.; Li, X. Prediction of Primary Dendrite Arm Spacing of the Inconel 718 Deposition Layer by Laser Cladding Based on a Multi-Scale Simulation. *Materials* **2023**, *16*, 3479. https://doi.org/10.3390/ma16093479

Academic Editor: Robert Pederson

Received: 20 March 2023
Revised: 13 April 2023
Accepted: 21 April 2023
Published: 29 April 2023

1. Introduction

Nickel-based superalloys are widely used as a key material in aircraft structures due to their high fracture resistance and excellent corrosion resistance at a high temperature [1]. However, repairing or replacing damaged superalloy components using conventional production processes can be expensive. Laser cladding technology offers a cost-effective alternative for component repair or rapid manufacturing [2]. Laser cladding technology involves complex physical processes, including solid–liquid phase change, fluid flow and latent heat of melting, and is influenced by various process parameters such as laser power, scanning velocity, powder feeding rate and laser beam diameter. The choice of parameters significantly impacts the final microstructure morphology and mechanical properties [3]. Among them, primary dendrite arm spacing (PDAS) is an important parameter for characterizing microstructure, and can influence the performance of components, especially in laser processing. Due to the high temperature gradient and rapid cooling in laser cladding, the microstructure of the alloy produced by this process differs from that produced by conventional directional solidification. Additionally, PDAS size can effectively control microsegregation, which is directly related to the formation of delta phase in nickel-based superalloys. During both forging and cladding processes, excessive phase aggregation is a major cause of cracking [4,5]. Therefore, it is helpful to understand the relationship between process parameters and PDAS.

The microstructure of the laser cladding layer is a topic of ongoing interest in the field of laser cladding [6–8]. Numerous studies have shown that the microstructure and mechanical properties of laser cladding parts are primarily controlled by the real-time thermal characteristics of molten pool and the corresponding solidification parameters (temperature gradient G and solidification rate R) [9]. Microstructure characteristics can be effectively predicted by associating process parameters with the real-time thermal characteristics and solidification parameters [10]. Conventional theoretical methods estimate PDAS using solidification parameters: $\lambda_1 = AG^m R^n$ [11], where A is the alloy constant and m and n are the model correlation indices. However, due to the complexity of the diffusion field around dendrites at high solidification rates, these approximate methods have large errors when applied to laser cladding [12].

Currently, high-temperature thermography and high-speed camera systems are used to monitor the solidification process of the molten pool [13–15] and measure the solidification parameters such as surface temperature and cooling rate. This allows analysis of the influence of solidification parameters on solidification structure size. However, on-site measurement of solidification conditions is challenging due to the high melt temperature and small size of the molten pool. Measured solidification parameters (such as temperature and fluid flow) are limited to the surface of the molten pool or thin melt layer, and experiment results are easily influenced by the working conditions [16–18].

Advances in computational ability and numerical methods have made computer simulation a popular tool for studying solidification problems. Many researchers have developed numerical models to analyze the evolution of solidification conditions in additive manufacturing processes and their impact on solidification microstructure characteristics [19,20]. The distribution of the macro-scale temperature field caused by the moving heat source is the key to determining solidification parameters. From this, solidification parameters such as the temperature gradient G and the solidification rate R at the S/L interface of the molten pool can be calculated. For micro-scale simulation, the cellular automata (CA) method and phase field (PF) method are commonly used to predict the solidification microstructure [21,22]. The phase-field method, in particular, is based on the thermodynamics of partial differential equations and explicitly tracks the interface, allowing the effective simulation of a complex S/L interface [23–25].

Currently, most multi-scale simulations of laser cladding use the finite element method with birth-death elements. However, this approach does not account for changes in the free surface or fluid flow in the molten pool and latent heat of melting. As a result, simulated solidification parameters may be inaccurate, leading to errors in predicting primary dendrite arm spacing in microscopic simulations [26].

The aim of this study is to link laser cladding process parameters with transient local thermal characteristics and solidification parameters to rapidly predict the solidification microstructure. Regarding Inconel 718 as the research object, a 3D transient laser cladding heat and mass transfer model was adopted to calculate the temperature gradient G and solidification rate R at the S/L interface of the laser cladding molten pool under different process parameters. Then, this was combined with a quantitative phase field model to further calculate the PDAS at the top of the molten pool. To validate the model, laser cladding experiments were carried out to analyze the integrated influence of process parameters on microstructure.

2. Mathematical Model

2.1. Macroscopic Heat and Mass Transfer Model

During the laser cladding process, the laser beam moves continuously and the solidification front keeps pace with the heat source, as shown in Figure 1. The macroscopic temperature distribution caused by the moving heat source is the key to determining the solidification parameters. A 3D macroscopic heat and mass transfer model is established to simulate the temperature field, velocity field and molten pool morphology. The temperature gradient G and solidification rate R along the normal line of the S/L interface are further calculated. As the interface for the solid phase grows, each point at the solidification front grows in a direction perpendicular to the S/L interface at the corresponding R rate. $R = V\cos\theta$, where θ is the angle between the laser scanning direction and the normal solidification interface direction of the solidification interface in terms of molten pool geometry.

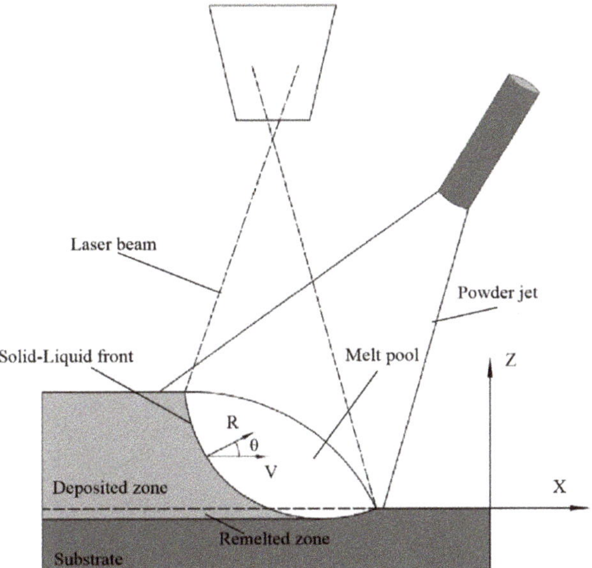

Figure 1. Schematic diagram of laser cladding.

The simplified assumptions are as follows [27,28]:

(1) Liquid fluid flow in the molten pool is Newtonian, laminar and incompressible;
(2) The mushy region is assumed as a porous medium with isotropic permeability;
(3) Heat flux from heated powder and heat loss due to evaporation are neglected;
(4) The effect of shielding gas, powder feeding gas and powder on the surface of the molten pool is ignored.

The continuity equation is:

$$\frac{\partial \rho}{\partial t} + \nabla \cdot (\rho \boldsymbol{u}) = 0 \tag{1}$$

Due to the existence of S/L phase change in the laser cladding, the influence of latent heat on melting should be considered in the heat transport equation; the energy equation is [29]:

$$\rho c_p \left(\frac{\partial T}{\partial t} + \boldsymbol{u} \cdot \nabla T \right) = \nabla \cdot (k \nabla T) - \frac{\partial H}{\partial t} - \rho \boldsymbol{u} \cdot \nabla H \tag{2}$$

In the above equation, the second and third terms on the right side represent unsteady terms of phase change enthalpy and the phase change enthalpy due to convection, respectively.

Where ρ is density, c_p is specific heat, T is temperature, t is time, k is heat conduction dilution, ΔH is melting latent enthalpy capacity, and $\Delta H = LF_l$, L is latent heat of melting. The liquid mass fraction F_l is defined as,

$$F_l = \begin{cases} 1, & T > T_l \\ \frac{T - T_s}{T_l - T_s}, & T_s < T \leq T_l \\ 0, & T < T_s \end{cases} \tag{3}$$

The incompressible Navier–Stokes equation describes the liquid metal in the molten pool. Since velocity in the solid phase region is 0, a source term in the momentum equation is introduced to suppress velocity field in the solid phase region. In this study, the time source term in the liquid phase region is set to 0, leaving the momentum equation unchanged. In the solid phase region, however, the source term is infinite, and the remaining terms of the momentum equation become infinitesimally small. In the transition region, the equation degenerates to Darcy's law relating porosity and pressure. Only pressure and source terms remain in the equation, while other terms become infinitesimally small.

$$\rho \frac{\partial u}{\partial t} + \rho(u \cdot \nabla)u = \nabla \cdot (\mu \nabla u) - \frac{\partial p}{\partial x} + K_0 \frac{(1 - F_l)}{F_l^3 + B} u \tag{4}$$

where μ is the viscosity, p is the pressure, and the third term on the right side of the Equation (4) represents the momentum dissipation term in the mushy region, which is based on the Carman–Kozeny Equation [30], K_0 is a constant related to the mushy region, and B is a very small number to prevent the denominator from being zero. Linear interpolation is used to interpolate the thermophysical parameters of the mushy region.

The laser heat flow input at the liquid/gas interface is as follows:

$$Q(x, y, t) = \frac{2P\eta_l}{\pi r_l^2} \exp\left(\frac{-2((x - X(t))^2 + (y - Y(t))^2)}{r_l^2}\right) - h_c(T - T_0) - \sigma_b \varepsilon(T^4 - T_0^4) \tag{5}$$

where P is the laser power, $X(t)$ and $Y(t)$ are the x and y coordinate values of the laser center with time, η_l is the absorption rate of the laser energy, r_l is the radius of the laser source, h_c is the heat transfer coefficient, σ_b is the Stefan–Boltzmann constant, ε is the emissivity, and T_0 is the ambient temperature.

The boundary condition of the momentum equation of the gas–liquid interface is [29],

$$F_{L/G} = \sigma n^* \kappa - \nabla_s T \frac{d\sigma}{dT} \tag{6}$$

The two terms in Equation (6) represent capillary force and thermal capillary force, respectively. σ is the surface tension, n^* is the surface normal, and κ is the surface curvature.

The description of the surface geometry of the molten pool is achieved through the moving mesh of the Lagrangian–Euler method (ALE). Two velocities are considered at the liquid/gas interface: fluid flow velocity and boundary movement velocity, caused by powder addition. They can be expressed as [31]:

$$V_{L/G} = un^* - V_p \cdot n^* \tag{7}$$

where u is the flow velocity of the fluid of the liquid–gas interface, V_p represents the moving velocity of the liquid–gas interface, and the calculation formula of V_p is

$$V_p(t, x) = \frac{2m_f \eta_m}{\rho_m \pi r_p^2} \exp\left(\frac{-2\left((x - X(t))^2 + (y - Y(t))^2\right)}{r_p^2}\right) \tag{8}$$

In the above formula, m_f is powder feeding rate, η_m is powder capture rate, ρ_m is powder density, r_p is the radius of the laser source, $X(t)$ and $Y(t)$ are the moving track of the laser head with time, respectively.

2.2. Microscopic Phase-Field Model

The phase field model adopts the quantitative alloy phase field model proposed by Echebarria et al. [32], which is established under the thin interface limit to eliminate the dependence of the interface thickness, so as to describe the (non-conserved) phase field ϕ and (conserved) the solute field c during the solidification of dilute binary alloys. An inverse solute diffusion term is introduced into the model to avoid the solute diffusion effect at the diffusion interface at lower solidification rates. Inconel 718 is simplified as a Ni-Nb binary system containing only FCC γ phase and liquid phase with a mass fraction of 5% Nb. The scalar phase field parameter ϕ indicates whether a point in the two-dimensional field is liquid ($\phi = -1$), solid ($\phi = 1$), or a solid–liquid interface ($-1 < \phi < 1$).

Ignoring the influence of latent heat effect, the freezing temperature approximation is adopted.

$$T = T_0 + G\,(z - Rt) \tag{9}$$

where T_0 ($z = 0$, $t = 0$) is a reference temperature, G is the temperature gradient along the Z direction, and R is the pulling velocity.

The governing equations of phase field and supersaturated concentration field in a two-dimensional system,

$$\tau_\phi(\hat{n}, z)\frac{\partial\phi}{\partial t} = W_0^2\left\{\nabla\cdot\left[\alpha_s(\hat{n})^2\nabla\phi\right] + \partial_x\left(|\nabla\phi|^2\alpha_s(\hat{n})\frac{\partial\alpha_s(\hat{n})}{\partial(\partial_x\phi)}\right) + \partial_z\left(|\nabla\phi|^2\alpha_s(\hat{n})\frac{\partial\alpha_s(\hat{n})}{\partial(\partial_z\phi)}\right)\right\}$$
$$+\phi - \phi^3 - \lambda\left(1 - \phi^2\right)^2\left(U + \frac{z - Rt}{l_T}\right) \tag{10}$$

$$\left(\frac{1+k}{2} - \frac{1-k}{2}\phi\right)\frac{\partial U}{\partial t} = \nabla\cdot\left[D_l\frac{(1-\phi)}{2}\nabla U + \vec{j}_{at}\right] + [1 + (1+k)U]\frac{1}{2}\frac{\partial\phi}{\partial t} \tag{11}$$

where $\tau_\phi(\hat{n}, z) = \tau_0(\alpha_s(\hat{n}))^2\left[1 - (1-k)\frac{(z - Rt)}{l_T}\right]$, which is the temperature-dependent relaxation time. $\alpha_s(\hat{n}) = (1 - 3\delta)\left[1 + \frac{4\delta}{1-3\delta}\left(\hat{n}_x^4 + \hat{n}_z^4\right)\right]$, which represents the two-dimensional fourfold anisotropy in the system, δ is strength of the surface tension anisotropy. $l_T = \frac{|m|c_\infty(1/k-1)}{G}$ is the thermal length, where $c_\infty \equiv c$ ($z = +\infty$) is an initial alloy concentration (wt.%) far from the solidification front, m is liquid line coefficient, k is equilibrium partition coefficient, D_l is the diffusivity of solute in liquid, $\hat{n} = \frac{\nabla\phi}{|\nabla\phi|}$, which is the unit vector normal to the interface. $U = \frac{1}{1-k}\left[\frac{ck/c_\infty}{(1-\phi)/2+k(1+\phi)/2} - 1\right]$, generalized supersaturation field. $\vec{j}_{at} = \frac{1}{2\sqrt{2}}W_0[1 + (1-k)U]\frac{\nabla\phi}{|\nabla\phi|}\frac{\partial\phi}{\partial t}$, anti-solute interception term. The anti-trapping current is given by Karma [33]. W_0 is the value of setting interface thickness. λ is used as a parameter for the numerical convergence of the control model. τ_0 stands for relaxation time. The interface thickness $W_0 = d_0\lambda/a_1$ and the relaxation time $\tau_0 = (d_0^2/D)a_2\lambda^3/a_1^2$ are interrelated from a thin interface analysis, where $a_1 = 0.8839$, $a_2 = 0.6267$, and λ is treated as a parameter to control the numerical convergence of the model.

2.3. Model Parameters

The high-temperature physical properties of the Inconel 718 were calculated using J-mat pro and are presented in Table 1. A continuous laser with a radius of 1.1 mm was used. The powder mass flow rate is 10 g/min, with a capture efficiency of 0.9 and a flow density of 8.19 g/cm^3. The powder flow radius was set to 2.3 mm. The values of these parameters are listed in Table 2.

Table 1. Physical parameters of Inconel 718.

Parameter	Value
Solidification spacing, ΔT	60 K
Melting point of pure nickel, T_m	1726.15 K
Inconel 718 liquidus, T_L (calculated by J-mat pro)	1635.14 K
Inconel 718 solidus, T_S (calculated by J-mat pro)	1489.63 K
Fluid Diffusion Coefficient, D (calculated by J-mat pro)	0.7×10^{-9} m^2/s
Gibbs-Thomson Coefficient, Γ	1.8×10^{-7} Km

Table 2. Macro simulation parameters.

Process Parameters	Value
Radius of beam, r_l	1.1 mm
Powder feeding rate, m_f	10 g/min
Powder flow radius, r_p	2.3 mm
Powder capture efficiency, η_m	0.9
Energy absorption efficiency of powder, η_p	0.22
Convective Heat Transfer Coefficient, h_c	10 W/(m^2·K)

The simulation model is a 10 mm × 20 mm × 5 mm cuboid with the cladding surface on the upper surface, serving as the input boundary for both the energy and mass transfer equations and the lifting surface for the ALE dynamic mesh. Mesh displacement is determined by the rise of the free surface due to powder addition and the influence of molten pool flow velocity on liquid surface advancement.

The phase field and concentration equations are solved on uniform meshes with a width of 50 μm in the x direction, and phase field parameters are listed in Table 3. The phase field governing equation ϕ is solved using a finite difference separation dispersion scheme, while the saturation concentration field U is solved using a finite-volume method with zero flux boundary conditions on both boundaries. The mesh spacing $\Delta x = \Delta y = 0.005$ μm, and the time step $\Delta t = 0.02\tau_0$, while the value of W_0 is about 10 times smaller than the dendrite tip radius calculated by the sharp interface model.

Table 3. Phase field parameters.

Parameter	Value
Initial alloy mass fraction, c_0	5%
Equilibrium Partition Coefficient, k_e	0.48
Liquidus Slope, m_l	-10.5 K%$^{-1}$
Equilibrium Freezing Range, T	57 K
Anisotropy Strength, δ	3%
Capillary Length, d_0	8.0×10^{-9} m
Liquid Diffusion Coefficient, D_l	3×10^{-9} m^2 s^{-1}
Solid Diffusion Coefficient, D_s	10^{-12} m^2 s^{-1}

The maximum phase field interface thickness $W_0 = 0.01$ μm, $\lambda = 1.377$ initializes the simulation along the bottom domain (at $y = 0$), and the S/L interface is randomly perturbed by small amplitude.

3. Experimental Materials and Scheme

Inconel 718 powder with a particle size of 45–105 μm and good flow characteristics was used in the experiment. The powder was dried at 100 °C for 2 h in a blast dryer prior to use. The elemental composition of powder elements is shown in Table 4:

Table 4. Inconel 718 powder element content (wt.%).

Element	Ni	Co	Mo	Al	Nb	C	Cr
Content	52	0.1	2.94	0.53	5.16	0.037	18.7

A CO_2 laser with a maximum power output of 4000 W was utilized in our experiment. Our team's previous research indicated that laser power below 1500 W resulted in partial melting of the track due to powder deposition. Conversely, laser power above 2300 W caused powder vaporization and substrate melting due to splashing. Energy accumulation led to powder vaporization and substrate melting at scanning speeds below 4 mm/s. At higher scanning speeds (12 mm/s), partial melting of the trajectory was observed due to insufficient energy to melt the powder. So, laser power (1500 W, 1900 W, 2300 W) and scanning velocity (4 mm/s, 8 mm/s, and 12 mm/s) were selected for the orthogonal experiment, with a shielding gas flow rate of 10 L/min. The substrate was a $20 \times 10 \times 5$ mm forging.

After wire cutting, surface oxide layers and stains were removed using an iron brush to prevent interference with cladding. Three cross sections perpendicular to the scanning direction were taken, and samples were ground and polished (using sandpaper grits of 240#, 600#, 800#, 1200#, and 2000# followed by polishing with W3.5 and W2.5 pastes). Samples were etched for 25 s using a solution of HCL, H_2SO_4 and $CuSO_4$.Geometrical morphology was observed using an optical metallographic microscope and analyzed.

This study focused on dendrite growth in the deposited zone rather than the remelting zone. To ensure consistency in point positions across different process parameters, dendrite arm spacing was measured in an area below 0.2 mm from the top and near the centerline. The statistical method is to select three random-shaped ranges in the target area, and the area of the shape as S, and the number of dendrites is N. (Regions with less than 50% coverage at the boundary were marked as 0 and those with more than 50% coverage were marked as 1).

Three sections were taken from each sample and three positions were randomly selected within the target area for each cross-section. The cladding process was found to be quasi-steady-state, with G and R remaining unchanged for different x-coordinate values as long as y and z positions were consistent. Data from different cross-sections under the same process parameters were used to increase accuracy.

4. Results and Discussion

4.1. Solidification Characteristics of Molten Pool

Nine cladding processes were simulated according to the orthogonal test. As energy is continuously input, the temperature in the molten pool increased until it exceeded the solidus temperature and a molten pool formed due to solid–liquid phase change. Figure 2a shows the temperature field at 0.5 s for a scanning velocity of $V = 8$ mm/s and laser power of $P = 1900$ W.

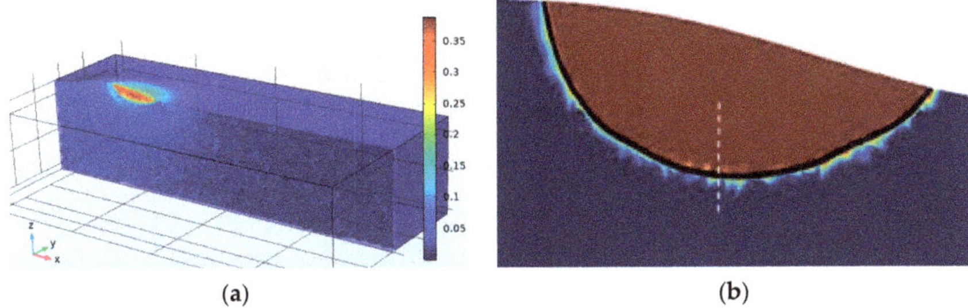

(a)   (b)

Figure 2. (a) The cross-section temperature field at 0.5 s, (b) Liquid phase temperature isotherm.

A quasi-steady-state is achieved after 0.5 s with an essentially unchanged isotherm shape. The back edge of the melt pool represented the solidification front and, after reaching quasi-steady-state, solidification parameters remained constant for all positions along the front. The longitudinal section of the molten pool (t = 1593 K) is taken as contour lines in Figure 2b, with the left part representing the solidification front.

As the laser source moves, a 3D solidification front surface also moves. Solidification parameters for all positions along the front are projected onto the cross-section of the deposition track to analyze molten pool characteristics (Figure 3).

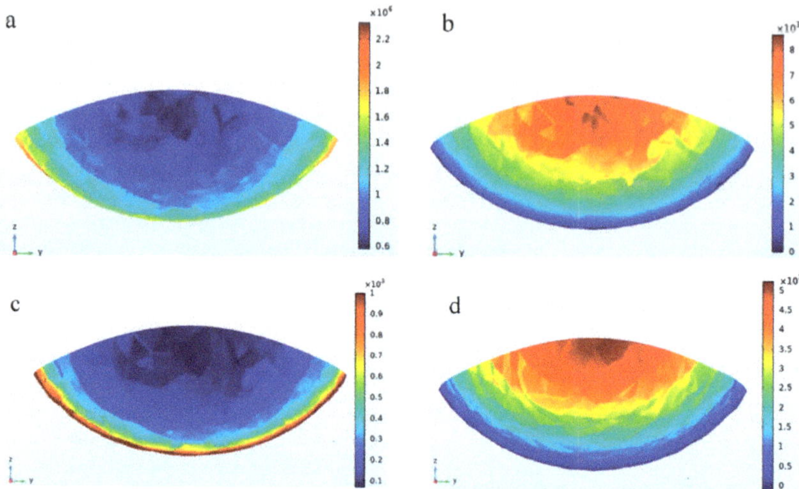

Figure 3. Computed contours of solidification characteristics: (**a**) temperature gradient G, (**b**) solidification growth rate R, (**c**) G/R, and (**d**) $G \times R$.

As shown in Figure 3a, G increases from 1.6×10^3 K/m along the periphery to 2.2×10^3 K/m at the center of the cross-section, before decreasing sharply in the middle of the molten pool, and the temperature gradient G is smallest at the top of the pool.

The R value is shown in Figure 3b, and it is 8 mm/s at the top of the molten pool, which is the same as the scanning velocity, and decreases gradually to 0.2 mm/s at the surface of the deposited track.

Figure 3c,d show the contour of G/R and $G \times R$ over the projective plane, respectively. Higher G/R and lower $G \times R$ are observed at the bottom of the solidification front, while lower G/R and higher $G \times R$ are observed at the top. The G/R ratio governs the solidification mode, while $G \times R$ controls the scale of the solidification microstructure. In this study, microstructure scale is coarser at the bottom of the deposited track, and finer at the top. Solidification mode may transform from planar to cellular, columnar dendritic, and equiaxed dendritic as G/R decreases.

The values of G were calculated for each direction, and compared with the cladding layer obtained by the microstructure experimentally. As shown in Figure 4, $G_{[001]}$ in the direction of [001] and $R_{[001]}$ are dominant in region A, while $G_{[100]}$ in the direction of [100] is dominant in region B, and $G_{[010]}$ in the direction of [010] is dominant in region C.

The simulation results agree well with the experimental results. Columnar crystals grew vertically from the bottom of the molten pool and the dendrites deflected at positions where $G_{[001]}$ is less than $G_{[100]}$, forming a cross shape in the cross-section.

Dendrite deflection is shown in Figure 5, with Figure 5a,b showing enlarged views of regions A and B in Figure 4, respectively, and Figure 5c showing a longitudinal section. A clear boundary line for dendrite growth direction can be observed in Figure 5c, with the dendrite growth direction at the top of the deposition area above the boundary differing from that at the bottom. Dendrites at the bottom of the deposition layer were thicker columnar crystals, while the microstructure at the top was more refined and smaller in size than dendrites at the bottom of the molten pool.

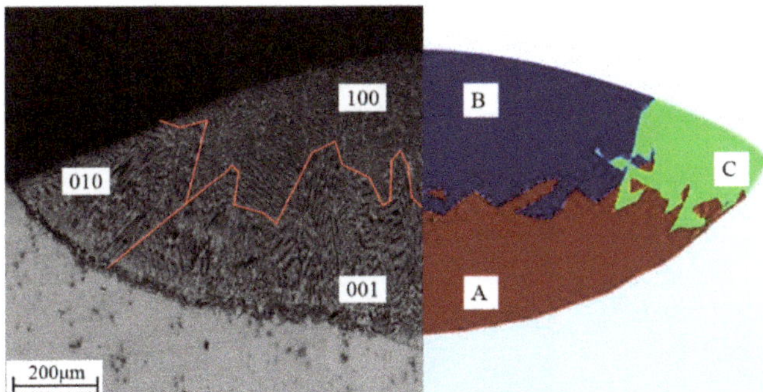

Figure 4. Comparison of dendrite growth direction between test and simulation; (A) $G_{[001]}$ dominant region; (B) $G_{[100]}$ dominant region; (C) $G_{[010]}$ dominant region.

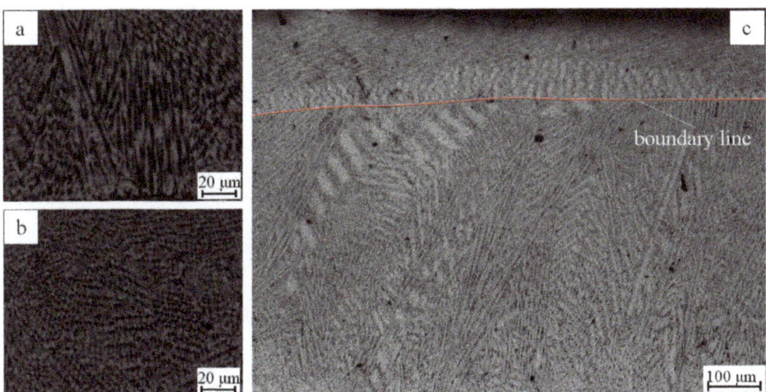

Figure 5. (**a**) Local enlarged drawing of Figure 4A; (**b**) Local enlarged drawing of Figure 4B; (**c**) the longitudinal section diagram.

In this study, the microstructure of nine process parameters is considered, and G and R values are calculated as the comparative data in Table 5.

Table 5. Nine process parameters and G, R values.

No.	P (W)	V (mm/s)	G (K/m)	R (m/s)
1	1500	4	4.42×10^5	1.47×10^{-3}
2	1500	8	6.09×10^5	3.21×10^{-3}
3	1500	12	6.48×10^5	6.23×10^{-3}
4	1900	4	4.65×10^5	1.68×10^{-3}
5	1900	8	5.41×10^5	3.95×10^{-3}
6	1900	12	5.69×10^5	6.38×10^{-3}
7	2300	4	4.76×10^5	1.77×10^{-3}
8	2300	8	4.85×10^5	4.03×10^{-3}
9	2300	12	4.98×10^5	7.80×10^{-3}

4.2. Simulation of Microstructure

G and R values for each process parameter were obtained and input into the phase field model. After the initial transients, Mullins–Sekerka instability rapidly developed at the solid–liquid interfaces, resulting in cellular structures that merged or split during growth. As solidification continue, the number of dendrites in the simulation region remain constant after a certain stage, with dendrite tips growing at a constant rate equal to the solidification rate, and average distance between neighboring γ-cells tips remaining constant. Steady-state dendrite morphology for various processes is shown in Figure 6. Mean PDAS is estimated from simulated cellular patterns by counting the number of cells along the system width (perpendicular to growth direction) and dividing by system width. The mean steady-state PDAS is between 5 μm to 12 μm in this case.

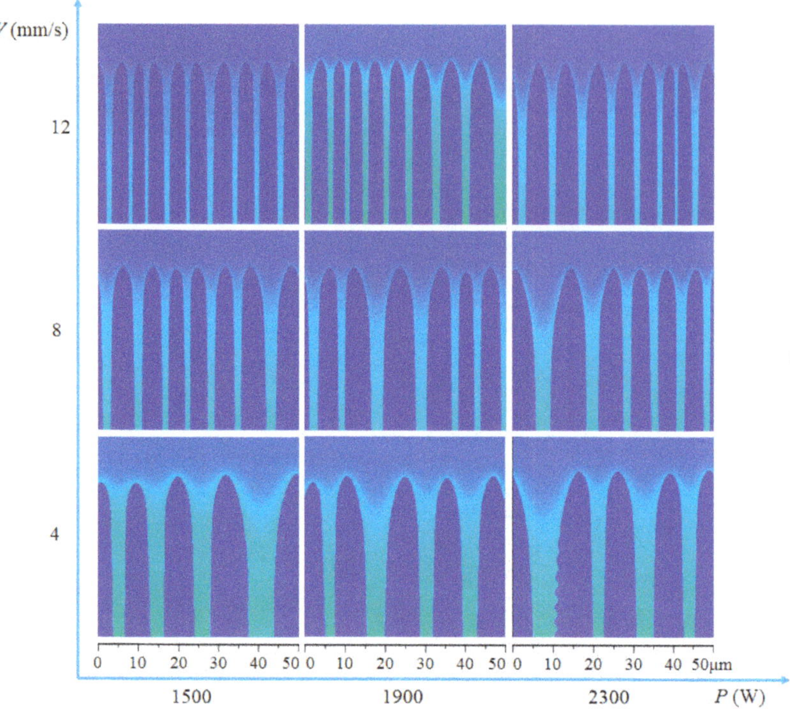

Figure 6. Steady-state cellular growth fronts for nine different solidification conditions.

Figure 6 illustrates significant variations in Nb composition in the liquid ahead of tips, between cells, and across solid cells. Growing cells reject niobium into the liquid, enriching intercellular regions. Solidification of Inconel 718 consistently results in the formation of an Nb-rich phase. Related test results indicate that niobium is concentrated in the delta phase region, which is relatively brittle and prone to cracking [5]. Thus, simplification as a Ni-Nb binary alloy is plausible for describing major solidification behavior.

4.3. Influence of the Processing Parameters on the PDAS

A comparison of the PDAS test and simulation results for Inconel 718 with nine process parameters is shown in Figure 7. Data comparison revealed a maximum error of 9.58% for process parameters 2300 W and 8 mm/s. Little difference is observed between PDAS values measured by simulation and experiment. The PF simulation was two-dimensional and solidification geometry may change slightly under solidification conditions in reality. However, the trend of simulation and experiment changing with scanning velocity and laser power was generally consistent, verifying model accuracy. Primary dendrite spacing decreased significantly with increasing laser scanning velocity and increased gently with increasing laser power.

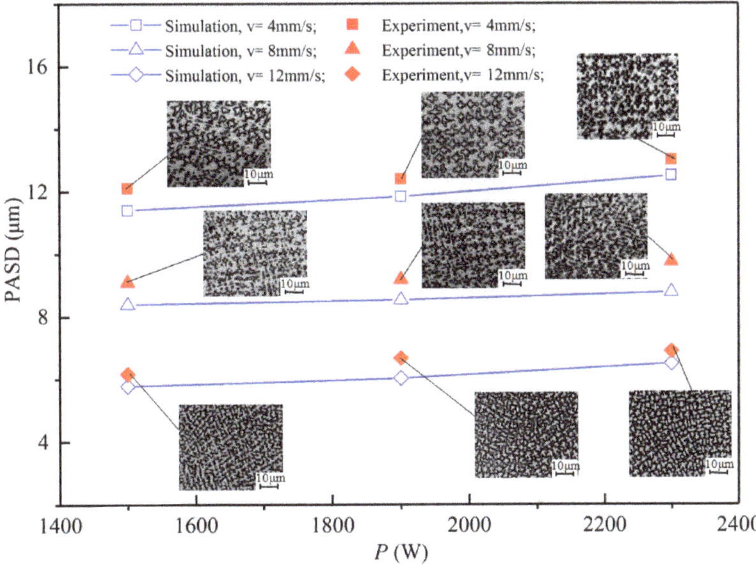

Figure 7. Variation trends of the PDAS with *P* at different *V*, OM image showing the transverse section of the laser track processed for nine different solidification conditions.

To analyze the trend of PDAS with process parameters, it is necessary to reveal the relationship between solidification conditions and process parameters as PDAS is directly determined by specific alloy solidification conditions. Higher cooling rates resulted in smaller dendrite spacing, consistent with simulation results showing that G decreased with increasing P for a given V. This trend is similar to previous studies and is due to decreased heat flux from larger pools produced by higher P. The effect of process parameters on solidification rate R is more complex, with R increasing with increasing P at lower power and higher scanning velocity but P having no significant effect at higher scanning velocity. These comparisons show that the selection of characteristic solidification conditions primarily affects PDAS size but does not significantly change PDAS evolution trends. As long as suitable solidification characteristic conditions are selected, PDAS does not change during solidification.

It can be seen from Figure 8 that when laser power increases from 1500 W to 2300 W at a scanning velocity of 4 mm/s, PDAS increases from 5.79 μm to 6.51 μm. Increasing laser power increased molten pool energy input and heat accumulation, decreasing the molten pool cooling rate and leading to dendrite structure coarsening.

Scanning velocity (V) had a greater influence on PDAS than power. At constant power, PDAS increases significantly with decreased scanning velocity. At 1500 W power, increasing the scanning velocity from 12 mm/s to 4 mm/s increases PDAS by 197%. Scanning velocity changes affect both laser-molten pool interaction time and molten pool energy input. Increasing the scanning velocity decreases laser energy input and interaction time, reducing molten pool heat accumulation and increasing the substrate quenching effect. This significantly increases the molten pool cooling rate and refines the dendrite structure. Scanning speed is the most readily controllable process parameter and can be adjusted in real-time to achieve a desired microstructure.

It is generally accepted that low-energy input refines the solidification microstructure. Decreasing the laser power or increasing the scanning velocity directly decreases line energy input, increasing the molten pool cooling rate and refining the dendrite structure. Scanning velocity plays a leading role in the molten pool cooling rate and solidification structure, consistent with results reported by Muvvala [15].

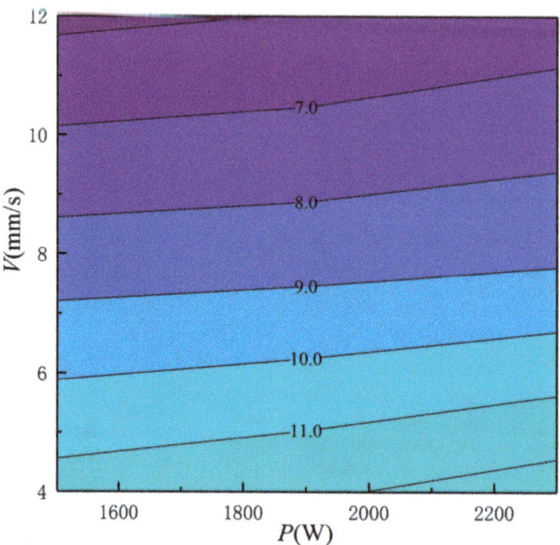

Figure 8. Combined effects of P, V on the PDAS (in μm).

5. Conclusions

In this study, a numerical model for predicting the PDAS of Inconel 718 laser cladding was established by combining a macroscopic transient 3D heat and mass transfer model with a quantitative phase field model. Solidification behavior in the molten pool under different process parameters was simulated, and the calculated results agreed well with the experimental values.

Simulated laser cladding showed that different positions within the cladding layer experienced different temperature histories and generated different dendrite morphologies. Finer equiaxed grains were obtained at the top of the cladding layer.

PDAS increased with increasing power and scanning velocity. Increased canning velocity resulted in an increase in monotonic PDAS with increasing power. Scanning velocity had a more pronounced effect on PDAS than power, with PDAS increasing significantly with increased scanning velocity. At 1500 W power, increasing the scanning speed from 12 mm/s to 4 mm/s increased PDAS by 197%. A finer microstructure could be obtained by combining a higher scanning speed with lower laser power.

The results showed that the multi-scale model provided better PDAS prediction for different processes, aiding the control and optimization of the laser cladding dendrite structure.

Author Contributions: Conceptualization, X.K.; Methodology, X.K.; Software, Z.J.; Validation, Z.J.; Formal analysis, L.M.; Data curation, J.D. and X.L.; Writing—original draft, Z.J.; Project administration, X.K. All authors have read and agreed to the published version of the manuscript.

Funding: This research was funded by the National Key Research and Development Program of China grant number SQ2019YFB1704500; the National Science and Technology Major Project grant number J2019-V-0009-0103; the State Ministry of Science and Technology Innovation Fund of China grant number 2018IM030200.

Conflicts of Interest: The authors declare no conflict of interest.

References

1. Babu, S.S.; Raghavan, N.; Raplee, J.; Foster, S.J.; Frederick, C.; Haines, M.; Dinwiddie, R.; Kirka, M.K.; Plotkowski, A.; Lee, Y.; et al. Additive manufacturing of nickel superalloys: Opportunities for innovation and challenges related to qualification. *Metall. Mater. Trans. A* **2018**, *49*, 3764–3780. [CrossRef]
2. Laeng, J.; Stewart, J.G.; Liou, F.W. Laser metal forming processes for rapid prototyping—A review. *Int. J. Prod. Res.* **2000**, *38*, 3973–3996. [CrossRef]

3. Wang, L.; Felicelli, S. Analysis of thermal phenomena in LENS™ deposition. *Mater. Sci. Eng. A* **2006**, *435*, 625–631. [CrossRef]

4. Azadian, S.; Wei, L.Y.; Warren, R. Delta phase precipitation in Inconel 718. *Mater. Charact.* **2004**, *53*, 7–16. [CrossRef]

5. Moshtaghi, M.; Safyari, M.; Mori, G. Combined thermal desorption spectroscopy, hydrogen visualization, HRTEM and EBSD investigation of a Ni–Fe–Cr alloy: The role of hydrogen trapping behavior in hydrogen-assisted fracture. *Mater. Sci. Eng. A* **2022**, *848*, 143428. [CrossRef]

6. Thijs, L.; Verhaeghe, F.; Craeghs, T.; Van Humbeeck, J.; Kruth, J.-P. A study of the microstructural evolution during selective laser melting of Ti–6Al–4V. *Acta Mater.* **2010**, *58*, 3303–3312. [CrossRef]

7. Kurz, W.; Trivedi, R. Rapid solidification processing and microstructure formation. *Mater. Sci. Eng. A* **1994**, *179*, 46–51. [CrossRef]

8. Amato, K.N.; Gaytan, S.M.; Murr, L.E.; Martinez, E.; Shindo, P.W.; Hernandez, J.; Collins, S.; Medina, F. Microstructures and mechanical behavior of Inconel 718 fabricated by selective laser melting. *Acta Mater.* **2012**, *60*, 2229–2239. [CrossRef]

9. Sames, W.J.; List, F.A.; Pannala, S.; Dehoff, R.R.; Babu, S.S. The metallurgy and processing science of metal additive manufacturing. *Int. Mater. Rev.* **2016**, *61*, 315–360. [CrossRef]

10. Collins, P.C.; Brice, D.A.; Samimi, P.; Ghamarian, I.; Fraser, H.L. Microstructural control of additively manufactured metallic materials. *Annu. Rev. Mater. Res.* **2016**, *46*, 63–91. [CrossRef]

11. Hunt, J.D. *Solidification and Casting of Metals*; The Metal Society: London, UK, 1979.

12. Ghosh, S.; Ma, L.; Ofori-Opoku, N.; Guyer, J.E. On the primary spacing and microsegregation of cellular dendrites in laser deposited Ni–Nb alloys. *Model. Simul. Mater. Sci. Eng.* **2017**, *25*, 065002. [CrossRef]

13. Khanzadeh, M.; Chowdhury, S.; Marufuzzaman, M.; Tschopp, M.A.; Bian, L. Porosity prediction: Supervised-learning of thermal history for direct laser deposition. *J. Manuf. Syst.* **2018**, *47*, 69–82. [CrossRef]

14. Xiao, H.; Li, S.; Han, X.; Mazumder, J.; Song, L. Laves phase control of Inconel 718 alloy using quasi-continuous-wave laser additive manufacturing. *Mater. Des.* **2017**, *122*, 330–339. [CrossRef]

15. Muvvala, G.; Karmakar, D.P.; Nath, A.K. Online monitoring of thermo-cycles and its correlation with microstructure in laser cladding of nickel based super alloy. *Opt. Lasers Eng.* **2017**, *88*, 139–152. [CrossRef]

16. Hofmeister, W.; Griffith, M. Solidification in direct metal deposition by LENS processing. *JOM* **2001**, *53*, 30–34. [CrossRef]

17. Xia, M.; Gu, D.; Yu, G.; Dai, D.; Chen, H.; Shi, Q. Influence of hatch spacing on heat and mass transfer, thermodynamics and laser processability during additive manufacturing of Inconel 718 alloy. *Int. J. Mach. Tools Manuf.* **2016**, *109*, 147–157. [CrossRef]

18. Liang, Y.J.; Wang, H.M. Origin of stray-grain formation and epitaxy loss at substrate during laser surface remelting of single-crystal nickel-base superalloys. *Mater. Des.* **2016**, *102*, 297–302. [CrossRef]

19. Gäumann, M.; Bezencon, C.; Canalis, P.; Kurz, W. Single-crystal laser deposition of superalloys: Processing–microstructure maps. *Acta Mater.* **2001**, *49*, 1051–1062. [CrossRef]

20. Acharya, R.; Sharon, J.A.; Staroselsky, A. Prediction of microstructure in laser powder bed fusion process. *Acta Mater.* **2017**, *124*, 360–371. [CrossRef]

21. Knapp, G.L.; Mukherjee, T.; Zuback, J.S.; Wei, H.L.; Palmer, T.A.; De, A.; DebRoy, T. Building blocks for a digital twin of additive manufacturing. *Acta Mater.* **2017**, *135*, 390–399. [CrossRef]

22. Boettinger, W.J.; Warren, J.A.; Beckermann, C.; Karma, A. Phase-field simulation of solidification. *Annu. Rev. Mater. Res.* **2002**, *32*, 163–194. [CrossRef]

23. Steinbach, I. Phase-field models in materials science. *Model. Simul. Mater. Sci. Eng.* **2009**, *17*, 073001. [CrossRef]

24. Moelans, N.; Blanpain, B.; Wollants, P. An introduction to phase-field modeling of microstructure evolution. *Calphad* **2008**, *32*, 268–294. [CrossRef]

25. Nie, P.; Ojo, O.A.; Li, Z. Numerical modeling of microstructure evolution during laser additive manufacturing of a nickel-based superalloy. *Acta Mater.* **2014**, *77*, 85–95. [CrossRef]

26. Keller, T.; Lindwall, G.; Ghosh, S.; Ma, L.; Lane, B.M.; Zhang, F.; Kattner, U.R.; Lass, E.A.; Heigel, J.C.; Idell, Y.; et al. Application of finite element, phase-field, and CALPHAD-based methods to additive manufacturing of Ni-based superalloys. *Acta Mater.* **2017**, *139*, 244–253. [CrossRef]

27. Qi, H.; Mazumder, J.; Ki, H. Numerical simulation of heat transfer and fluid flow in coaxial laser cladding process for direct metal deposition. *J. Appl. Phys.* **2006**, *100*, 024903. [CrossRef]

28. He, X.; Song, L.; Yu, G.; Mazumder, J. Solute transport and composition profile during direct metal deposition with coaxial powder injection. *Appl. Surf. Sci.* **2011**, *258*, 898–907. [CrossRef]

29. He, X.; Fuerschbach, P.W.; DebRoy, T. Heat transfer and fluid flow during laser spot welding of 304 stainless steel. *J. Phys. D Appl. Phys.* **2003**, *36*, 1388. [CrossRef]

30. Asai, S.; Muchi, I. Theoretical analysis and model experiments on the formation mechanism of channel-type segregation. *Trans. Iron Steel Inst. Jpn.* **1978**, *18*, 90–98. [CrossRef]

31. Morville, S.; Carin, M.; Peyre, P.; Gharbi, M.; Carron, D.; Le Masson, P.; Fabbro, R. 2D longitudinal modeling of heat transfer and fluid flow during multilayered direct laser metal deposition process. *J. Laser Appl.* **2012**, *24*, 032008. [CrossRef]

32. Echebarria, B.; Folch, R.; Karma, A.; Plapp, M. Quantitative phase-field model of alloy solidification. *Phys. Rev. E* **2004**, *70*, 061604. [CrossRef] [PubMed]

33. Karma, A. Phase-field formulation for quantitative modeling of alloy solidification. *Phys. Rev. Lett.* **2001**, *87*, 115701. [CrossRef] [PubMed]

Article

Experimental Study on Carbon Fiber-Reinforced Composites Cutting with Nanosecond Laser

Jihao Xu [1], Chenghu Jing [1], Junke Jiao [1,*], Shengyuan Sun [1], Liyuan Sheng [2,*], Yuanming Zhang [3], Hongbo Xia [1] and Kun Zeng [4]

1   School of Mechanical Engineering, Yangzhou University, Yangzhou 225009, China
2   PKU-HKUST ShenZhen-HongKong Institution, Shenzhen 518057, China
3   School of Mechanical and Vehicle Engineering, Linyi University, Linyi 276005, China
4   Yangzhou Hanjiang Yangzi Automobile Interior Decoration Co., Ltd., Yangzhou 225009, China
*   Correspondence: jiaojunke@yzu.edu.cn (J.J.); lysheng@yeah.net (L.S.)

Abstract: The carbon fiber-reinforced composite (CFRP) has the properties of a high specific strength, low density and excellent corrosion resistance; it has been widely used in aerospace and automobile lightweight manufacturing as an important material. To improve the CFRP cutting quality in the manufacturing process, a nanosecond laser with a wavelength of 532 nm was applied to cut holes with a 2-mm-thick CFRP plate by using laser rotational cutting technology. The influence of different parameters on the heat-affected zone, the cutting surface roughness and the hole taper was explored, and the cutting process parameters were optimized. With the optimized cutting parameters, the minimum value of the heat-affected zone, the cutting surface roughness and the hole taper can be obtained, which are 71.7 µm, 2.68 µm and 0.64°, respectively.

Keywords: carbon fiber reinforced composites; nanosecond laser; laser cutting; heat affected zone; surface roughness; hole taper

Citation: Xu, J.; Jing, C.; Jiao, J.; Sun, S.; Sheng, L.; Zhang, Y.; Xia, H.; Zeng, K. Experimental Study on Carbon Fiber-Reinforced Composites Cutting with Nanosecond Laser. *Materials* **2022**, *15*, 6686. https://doi.org/10.3390/ma15196686

Academic Editors: Anhai Li, Xiaoxiao Chen and Ya'ou Zhang

Received: 25 August 2022
Accepted: 20 September 2022
Published: 27 September 2022

Publisher's Note: MDPI stays neutral with regard to jurisdictional claims in published maps and institutional affiliations.

1. Introduction

Carbon fiber-reinforced composite (CFRP) has the advantages of a high specific strength, low density and light weight, and has become an important material for aerospace and automobile lightweight manufacturing [1,2]. Due to the great difference in physical properties between carbon fibers and the resin matrix in CFRP, problems such as delamination, fiber pull-out and tool wearing often occur when using traditional mechanical machining methods to cut CFRP. As a non-contact advanced processing technology, laser machining technology shows great potential in composite materials' cutting [3]. However, due to the thermal machining process, the ablation of resin matrix materials easily occurs during laser cutting, leading to problems such as a heat-affected zone (HAZ), delamination and fiber pull-out, which seriously affect the processing quality of CFRP [4,5]. In order to improve the quality of CFRP laser cutting, a large number of investigations have been carried out on CFRP laser cutting in recent years.

Jiang cut a 5-mm-thick CFRP plate with lasers, and he found that the HAZ gradually increased with an increase of power and that the HAZ reached the minimum value when the laser power was 60 W. He also found that the HAZ decreased with an increase of the laser scanning speed and that it reached the minimum value when the scanning speed was 10 m/s [6]. Ye used a 532 nm wavelength laser to make holes in CFRP, and compared the effects of cross filling, parallel filling and rotary scanning on the machining quality. It was found that the cutting efficiency of laser rotary cutting was significantly higher than that of the other two methods and that the HAZ was smaller [7]. Song found that adding carbon black particles in CFRP can significantly improve the cutting effect and reduce the occurrence of cracks and delamination [8]. Hua used a 500 W millisecond pulse laser to cut CFRP underwater. It was found that the HAZ was greatly reduced when compared with that cutting in air [9]. Zhang used lasers with different wavelengths to make holes in CFRP plates. It was found that the shorter the laser wavelength was, the higher the cutting surface quality and the smaller the HAZ that could be obtained [10]. Bluemel found that a continuous wave laser has a higher cutting speed than a pulsed laser but that the nanosecond laser has a narrower slit and higher tensile strength [11]. Negarestani found that when using mixed gas containing oxygen and inert gas as the auxiliary gas in CFRP cutting, the HAZ could be greatly reduced [12]. Li found that the cutting speed is an important factor affecting the HAZ, while the laser processing parameters have a limited influence on the surface strain distribution during the tensile load [13]. Lau found that the cutting direction of the laser affected the width of the HAZ and the cutting depth [14]. Jose found that cutting CFRP with a high repetition rate resulted in a very short cooling time and a large HAZ. To this end, a higher pulse duration and lower repetition frequency could reduce the HAZ [15].

As mentioned above, parameters such as the wavelength, the frequency, the laser mode (pulsed or continuous), the laser power and the cutting speed will influence the cutting quality of CFRP. Compared with the continuous laser, the short-pulsed laser can reduce the cutting HAZ and improve the cutting accuracy. Compared with the infrared laser, a laser with a short wavelength has a smaller heat input and smaller taper when cutting CFRP. Therefore, using a short-wavelength and short-pulsed laser to cut composites has become a trend to improve the cutting quality of CFRP. However, relations between the cutting quality, such as HAZ, surface roughness and the taper of the cutting hole, and the pulsed laser parameters, such as laser power and laser scanning speed, are not clear. It necessary to investigate them and make conclusions on the technical process, which is helpful in getting a high CFRP cutting quality with short-pulsed lasers in industry applications. To this end, in this paper, a nanosecond laser with a wavelength of 532 nm was applied to cut holes in a 2-mm-thick CFRP plate. Twenty-five groups of experiments with different cutting parameters were designed. The HAZ and the cutting surface roughness was measured by using SEM and a laser confocal microscope, and the taper of the cutting hole was calculated. The influence of different parameters (the laser power and the laser scanning speed) on the HAZ, the surface roughness and the hole taper was explored, and the cutting parameters were optimized, with the aim of laying a technological foundation for high-quality CFRP laser cutting.

2. Materials and Methods

The size of CFRP is 2 mm × 100 mm × 100 mm, and the volume contents of the carbon fibers and the epoxy resin are 68% and 32%, respectively. The fibers in the CFRP are continuous fiber, and the diameter of the fiber is about 5 μm. The CFRP was produced by Ningbo Institute of Materials technology and engineering. Additionally, the thermos-physical parameters of CFRP are shown in Table 1 [16].

Table 1. The thermos-physical parameters of CFRP.

Material	Thermal Conductivity ($W \cdot m^{-1}\,{}^\circ C^{-1}$)	Specific Heat ($J \cdot kg^{-1}\cdot{}^\circ C^{-1}$)	Vaporization Temperature ($^\circ C$)	Density (g/cm^3)
Epoxy resin	0.2	550	400	1.2
Carbon fibers	10.5	795.5	3300	1.76

A nanosecond laser is used to cut CFRP in this work. The laser wavelength is 532 nm, the pulse duration is 150 ns, the maximum power is 35 W, and the repetition frequency can be changed in the range of 50 KHz to 200 kHz. The parameters that can be set by the machining system include the scanning speed, the laser power and the repetition frequency of the nanosecond laser. The hole-cutting path and the cutting speed can be set by the scanning galvanometer. Figure 1a is a diagram of the nanosecond laser machining system. In the previous study, it was found that the cutting quality and efficiency of laser rotary cutting were higher than those of cross filling and parallel filling [16]. Therefore, the laser rotary cutting technique was used to cut CFRP in this work (Figure 1b). Furthermore, the spacing d was set at 0.1 mm and the radius r was set at 0.1 mm, and the radius of the cutting hole R was 2.5 mm, which was optimized in our previous work [16].

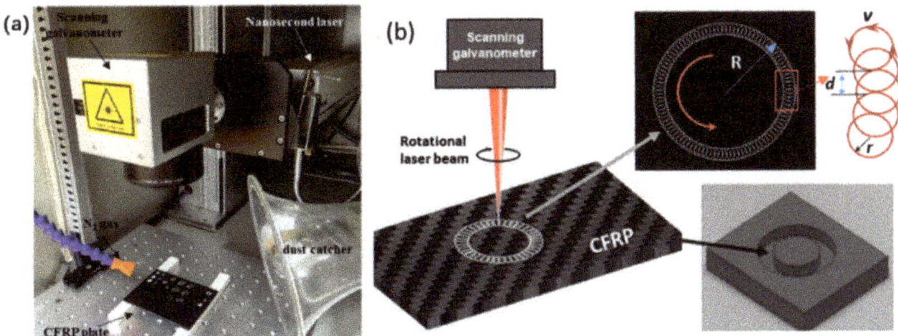

Figure 1. (**a**) The nanosecond laser machining system; (**b**) cutting CFRP with the laser rotary cutting technique [16].

In our previous research, it was found that the amount of energy (E) absorbed per unit area per unit time was determined by the laser power P, the laser scanning speed V and the laser spot diameter D, which is shown in Equation (1) [17]:

$$E = A \times \frac{P}{V \times D^2} \tag{1}$$

where A is the constant coefficient, which is determined by the laser absorption coefficient of the CFRTS.

In order to study the influence of different scanning speeds V and laser powers P on the CFRP machining quality, an orthogonal experiment was designed in this paper. The pulse duration was 150 ns, the repetition frequency was 60 KHZ, the laser power P was set at 21 W, 24.5 W, 28 W, 31.5 W and 35 W, and the cutting speed V was set at 800 mm/s, 900 mm/s, 1000 mm/s, 1100 mm/s and 1200 mm/s, respectively. A total of 25 groups of experiments were conducted, as shown in Table 2. During laser processing, nitrogen with a purity of 99.99% is used as the protective gas, and the nitrogen flow rate is about 8 L/min. Nitrogen can play the role of cooling the CFRP plate, reducing the generation of the HAZ, and blowing away the slag to improve the cutting efficiency. In order to observe the roughness of the cutting surface under different parameters, the CFRP plate was cut into 5-mm-wide strips. After cutting, the machining debris were removed by ultrasonic cleaning. During the measurement, three different positions on the cutting surface were measured, and the average value was calculated. The HAZ was measured by using a microscope, as shown in Figure 2a. From this figure, it can be seen that there is a region around the hole whose color is different from the CFRP matrix, and the width of this region is defined as the HAZ. The surface roughness of the CFRP cutting profile was detected by using the laser confocal microscope, as shown in Figure 2b. In this work, the size of the scanned area for the roughness measurement was 1054 μm × 1406 μm, and the average of Ra was chosen as the value of the surface roughness. After laser cutting, the microstructure of the cutting profile, such as fiber fractures and interlaminar tears, was detected by using SEM.

Table 2. Experimental design of CFRP laser cutting.

Parameter	21 W	24.5 W	28 W	31.5 W	35 W
800 mm/s	Sample 1-1	Sample 1-2	Sample 1-3	Sample 1-4	Sample 1-5
900 mm/s	Sample 2-1	Sample 2-2	Sample 2-3	Sample 2-4	Sample 2-5
1000 mm/s	Sample 3-1	Sample 3-2	Sample 3-3	Sample 3-4	Sample 3-5
1100 mm/s	Sample 4-1	Sample 4-2	Sample 4-3	Sample 4-4	Sample 4-5
1200 mm/s	Sample 5-1	Sample 5-2	Sample 5-3	Sample 5-4	Sample 5-5

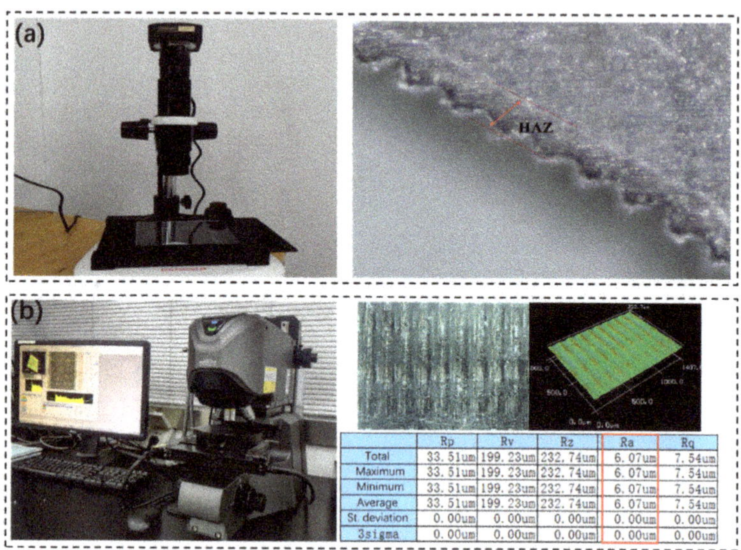

	Rp	Rv	Rz	Ra	Rq
Total	33.51um	199.23um	232.74um	6.07um	7.54um
Maximum	33.51um	199.23um	232.74um	6.07um	7.54um
Minimum	33.51um	199.23um	232.74um	6.07um	7.54um
Average	33.51um	199.23um	232.74um	6.07um	7.54um
St. deviation	0.00um	0.00um	0.00um	0.00um	0.00um
3sigma	0.00um	0.00um	0.00um	0.00um	0.00um

Figure 2. (**a**) Measuring the HAZ around the cutting hole with the microscope; (**b**) measuring the surface roughness with the laser confocal microscope.

The laser will produce a certain divergence angle during the actual operation. Therefore, there will be a certain taper during laser processing, as shown in Figure 3. In this experiment, the average value and standard deviation of the three tapers are calculated by three measurements at the entrance and exit of the beam. The taper θ can be calculated by the following formula:

$$\theta = arctan \frac{D_1 - D_2}{2h} \times \frac{180°}{\pi} \tag{2}$$

where D_1 is the beam entrance size, D_2 is the beam exit size, and h is the thickness of the CFRP plate.

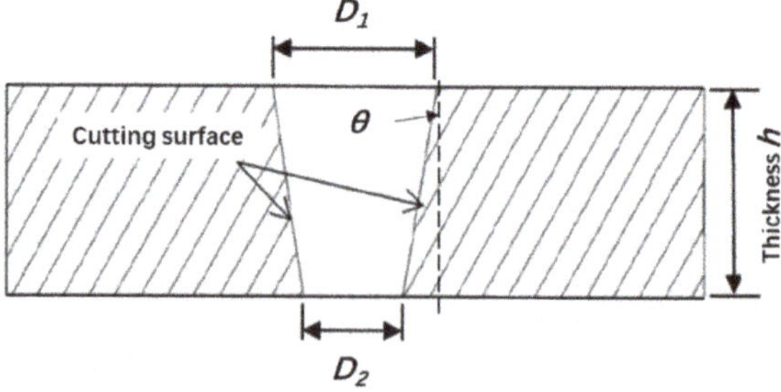

Figure 3. Hole taper during CFRP laser cutting.

3. Results and Discussions

3.1. Cutting Holes in CFRP Plate with Nanosecond Lasers

The process parameters of CFRP laser rotary cutting include the rotation diameter, the scanning space and the scanning speed. The focal length compensation speed of the z-axis is directly related to the above parameters. The focal length compensation speed should be matched with the material removal speed. The material removal ability with the negative defocus will become weak, or there will even be no removal ability. If the focal length compensation speed is too slow, it will affect the removal speed. Appropriate mathematical models and multiple experiments should be established to determine the appropriate calculation formula for the focal length compensation speed [7]. Finally, it is determined that the focal length of the 2-mm-thick plate is 0.5 mm per minute and that the time needed to punch through the hole is about 4 min. The holes of Sample 1-1 to Sample 5-5, cut with the parameters shown in Table 2, are shown in Figure 4a, and the setting diameter of the holes is 5 mm. As shown in Figure 4b,c, the inlet diameter is about 5.06 mm and the outlet diameter is about 4.96 mm when the laser power is 28 W, and the scanning speed is 1200 mm/s (Sample 5-3). The hole taper can be calculated by using the formula (2), as mentioned above, which is about 0.71°.

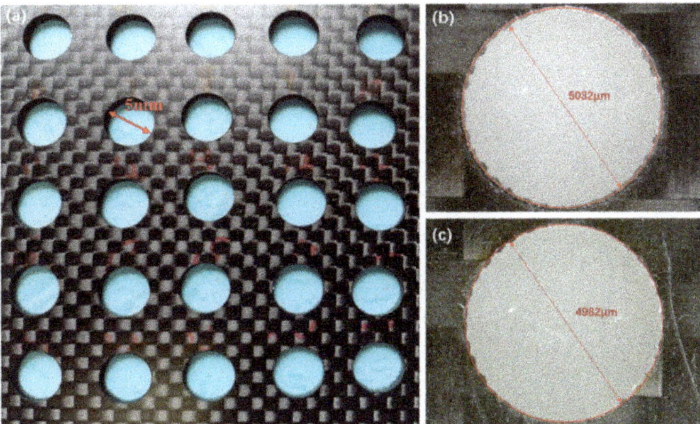

Figure 4. (**a**) Cutting holes in CFRP plate with nanosecond lasers for Sample 1-1 to Sample 5-5; (**b**) inlet of the CFRP laser cutting hole of sample 5-3; (**c**) outlet of the CFRP laser cutting hole of sample 5-3.

To clarify the HAZ in the CFRP matrix during the laser cutting process, the HAZ around the hole and the microstructure of the cutting profile were detected. Figure 5 shows the HAZ and the microstructure on the CFRP hole cutting surface for Sample 5-3. From Figure 5b, it can be seen that there is a circa 140-μm-wide HAZ around the inlet edge. With the laser interaction, the CFRP material is removed from the hole inlet during the cutting process, and the plasma produced during the cutting process goes away from the hole. Because of the high plasma temperature, the hole surface is heated and ablated, and a HAZ is generated. From Figure 5b, it can be seen that there are serrations around the hole inlet, which is caused by the excessive rotation space d of the rotating laser beam, as shown in Figure 1b. These serrations can be avoided by reducing the rotation space d. After cutting holes with the nanosecond laser, the CFRP hole is cut along the center line by lasers, and the hole profile is shown in Figure 5c,d. From this figure, it can be seen that the CFRP surface for nanosecond laser cutting is smooth, and there is no delamination and fiber pull-out on the cutting surface.

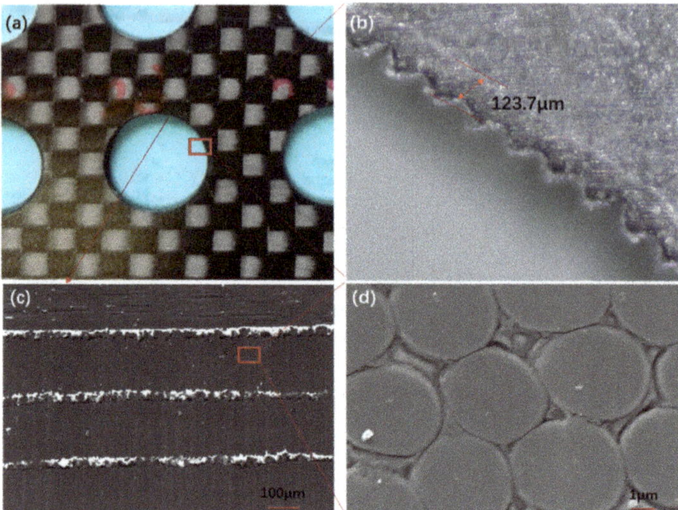

Figure 5. (**a**) Cutting hole in CFRP plate; (**b**) HAZ around the cutting hole; (**c**) cutting profile; (**d**) microstructure of the cutting profile.

The CFRP materials are vaporized due to the high energy of the pulsed nanosecond laser, and hence there is no mechanical force and impact force during the cutting process. There is no carbon fiber drawing and tearing compared with the mechanical cutting method. The HAZ can be controlled well, the cutting surface has a low roughness, and the taper of the cutting hole is about 0.71°. With this CFRP laser cutting technique and the parameters of Sample 5-3, holes with different shapes such as triangle and square can be drilled by changing the laser beam path (Figure 6). It can be seen that the cutting profile is clean and that there are no burrs and tears on the cutting surface.

Figure 6. CFRP laser cutting holes with different shapes.

3.2. CFRP Cutting Quality with Different Laser Parameters

The HAZ is the key factor affecting the cutting quality of CFRP. Figure 7 shows the HAZ of the hole edge under different powers and cutting speeds. It can be seen from the figure that when the laser power is 21 W, the HAZ first decreases and then gradually increases with the increasing of the laser scanning speed. When the scanning speed is 900 mm/s, the HAZ decreases first and then increases when increasing the laser power, and the minimum HAZ value of about 71.7 μm can be obtained when the laser power is 28 W. For other scanning speeds such as 800 mm/s, 1000 mm/s, 1100 mm/s and 1200 mm/s, nearly the same trend can be obtained, as shown in Figure 7.

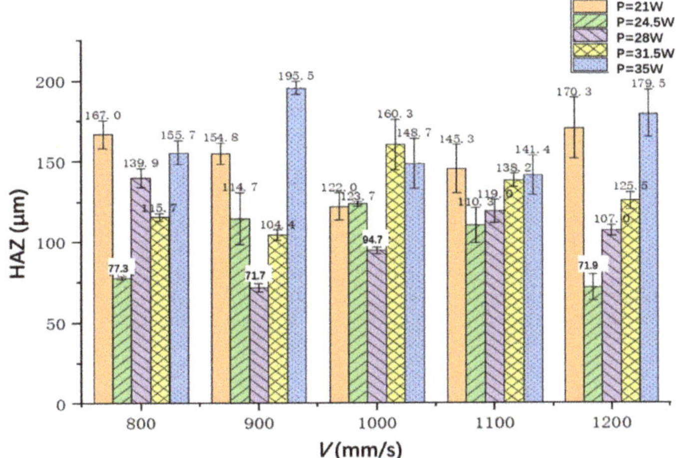

Figure 7. The HAZ with different laser powers and scanning speeds.

The minimum value is 122 μm at a speed of 1000 mm/s (Sample 3-1), and the maximum value is 170.3 μm at a speed of 1200 mm/s (Sample 5-1). When the laser power is 24.5 W, the HAZ increases first and then decreases when increasing the scanning speed, and the minimum value is 71.9 μm at a speed of 1200 mm/s (Sample 5-2). When the laser power is 28 W, the minimum value of HAZ is 71.7 μm at a scanning speed of 900 mm/s (Sample 2-3). In the experiment, the rotation radius and spacing of laser cutting can be reduced to reduce the serration of cutting. Increasing the flow of nitrogen shielding gas to cool the CFRP plate in a timely way could reduce the HAZ.

The surface roughness was measured by using a laser confocal microscope, and the roughness was kept within 12 um for these 25 cutting samples, as shown in Figure 8. From this figure, it can be seen that for a specific laser scanning speed, the cutting surface roughness increases first and then decreases when increasing the laser power. The minimum cutting profile roughness can be obtained when the laser power is 35 W and the scanning speed is 900 mm/s (Sample 2-5), which is about 2.68 μm. From Figure 8, it also can be seen that there is little change in the surface roughness with the change of the laser scanning speed when the laser power is 35 W, and the minimum value can be obtained when the speed is 900 mm/s. This means that a cutting profile with a low roughness can be obtained when the laser power is 35 W.

On the other hand, there is no obvious law between the roughness and the laser scanning speed. However, the minimum value roughness can be obtained under different laser scanning speeds, such as a minimum roughness of 3.08 μm when the laser power is 21 W and the scanning speed is 800 mm/s (Sample 1-1), 3.22 μm when the power is 35 W and the scanning speed is 1000 mm/s (Sample 3-5), 3.66 um when the power is 35 W and the scanning speed is 1100 mm/s (Sample 4-5), and 5.06 μm when the laser power is 35 W and the scanning speed is 1200 mm/s (Sample 5-5).

The hole taper can be calculated by measuring the diameter of the inlet and outlet of the hole, as shown in formula (1). From Figure 9, it can be seen that the hole taper can be controlled within 1.6° in the experiment. For a specific laser scanning speed, the hole taper increases first and then decreases when increasing the laser power. When the scanning speeds are 800 mm/s and 1200 mm/s, it can be seen that the hole taper decreases first and then increases when increasing the laser power, and the minimum values, which are about 0.75° and 0.71°, respectively, can be obtained when the laser power is 28 W. The minimum taper can be obtained when the laser power is 24.5 W and the scanning speed is 1100 mm/s (Sample 3-2), and it is about 0.64°.

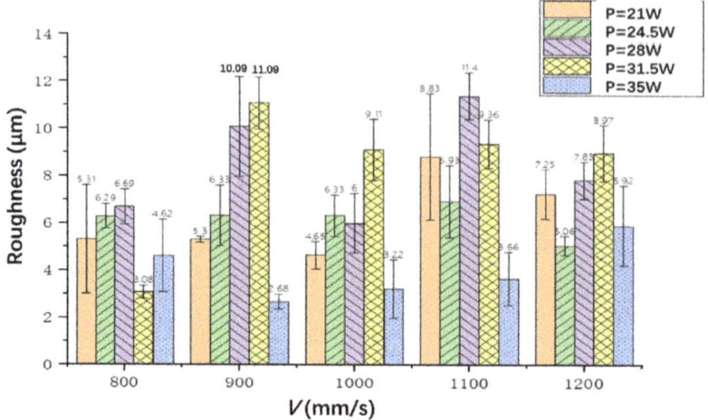

Figure 8. The surface roughness with different laser powers and scanning speeds.

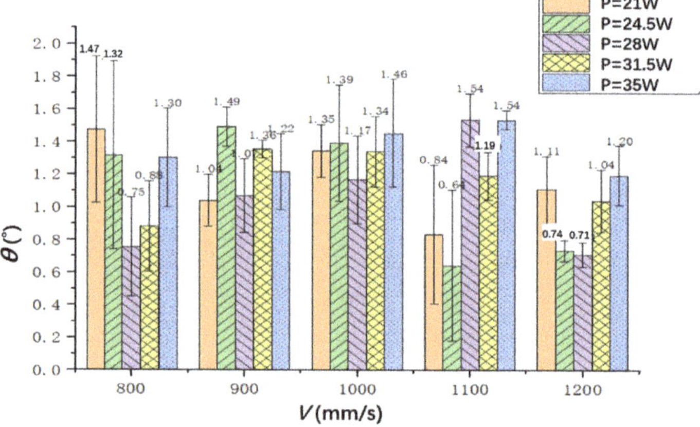

Figure 9. The hole taper with different laser powers and scanning speeds.

From this figure, it also can be seen that the difference between the minimum hole taper under different powers is small, at 0.84°, 0.64°, 0.71°, 0.88° and 1.20°, when the laser power is 21 W (Sample 4-1), 24.5 W (Sample 4-2), 28 W (Sample 5-3), 31.5 W (Sample 5-4) and 35 W (Sample 5-5), respectively. To this end, it can be concluded that the hole taper is less affected by laser power and that the taper is relatively stable. In later experiments, the focal length of the laser can be further controlled to keep the focal length on the cutting surface, which can further reduce the taper.

4. Conclusions

In this paper, a 532-nm-wavelength nanosecond laser was used to cut a 2-mm-thick CFRP plate. The effects of different laser powers and cutting speeds on the HAZ, surface roughness and taper are investigated. The following conclusions can be obtained.

(1) Laser rotational cutting technique can improve the efficiency and quality of cutting, but there will be small serrations, which can reduce the radius and spacing of laser rotary cutting and reduce the serration of cutting.

(2) Together, the laser power and the laser scanning speed affect the HAZ. Additionally, the amount of energy absorbed per unit area per unit time is determined by both of

them. When the power is 28 W and the scanning speed is 900 mm/s, the minimum value of the heat-affected zone can be obtained, and it is about 71.7 μm.

(3)   The cutting surface roughness increases first and then decreases when increasing the laser power for a specific laser scanning speed. The minimum value is 2.68 μm when the scanning speed is 900 mm/s and the laser power is 35 W.

(4)   The hole taper increases first and then decreases when increasing the laser power for a specific laser scanning speed. The minimum taper is 0.64° when the laser power is 24.5 W and the scanning speed is 1100 mm/s.

Author Contributions: Conceptualization, J.J., S.S., Y.Z. and K.Z.; methodology, C.J., J.J., L.S., Y.Z. and K.Z.; validation, J.X. and K.Z.; formal analysis, H.X.; investigation, S.S.; resources, J.X., C.J. and H.X.; data curation, C.J., S.S. and H.X.; writing—original draft preparation, J.X.; writing—review and editing, J.J.; visualization, J.X., J.J. and Y.Z.; supervision, L.S.; project administration, L.S.; funding acquisition, J.J. and L.S. All authors have read and agreed to the published version of the manuscript.

Funding: The authors gratefully acknowledge the support provided by Shenzhen Basic Research projects (JCYJ20200109144604020, JCYJ20200109144608205 and JCYJ20210324120001003) and Yangzhou Hanjiang Science and Technology project (HJZ2021003).

Institutional Review Board Statement: Not applicable.

Informed Consent Statement: Not applicable.

Data Availability Statement: Data sharing not applicable.

Conflicts of Interest: The authors declare no conflict of interest.

Abbreviations

P	laser power
V	laser scanning speed
E	the amount of energy absorbed per unit area per unit time
A	the absorption coefficient
D	the diameter of the laser spot
D_1	the diameter of the hole in the inlet
D_2	the diameter of the hole in the outlet
h	the thickness of CFRP plate
d	the laser beam rotation distance
r	the laser high-speed rotation radius
R	the hole radius
θ	the taper of the cutting hole
HAZ	the heat-affected zone
Ra	the cutting surface roughness

References

1.   Jiao, J.; Zou, Q.; Ye, Y. Carbon fiber reinforced thermoplastic composites and TC4 alloy laser assisted joining with the metal surface laser plastic-covered method. *Compos. Part B* **2021**, *213*, 108738. [CrossRef]

2.   Ye, Y.; Zou, Q.; Xiao, Y.; Jiao, J.; Du, B.; Liu, Y.; Sheng, L. Effect of Interface Pretreatment of Al Alloy on Bonding Strength of the Laser Joined Al/CFRTP Butt Joint. *Micromachines* **2020**, *12*, 179. [CrossRef] [PubMed]

3.   Oh, S.; Lee, I.; Park, Y.; Kim, H. Investigation of cut quality in fiber laser cutting of CFRP. *Opt. Laser Technol.* **2019**, *113*, 129–140. [CrossRef]

4.   Fischer, F.; Romoli, L.; Kling, R. Laser-based repair of carbon fiber reinforced plastics. *CIRP Ann. Manuf. Technol.* **2010**, *59*, 203–206. [CrossRef]

5.   Herzog, D.; Jaeschke, P.; Meier, O.; Haferkamp, H. Investigations on the thermal effect caused by laser cutting with respect to static strength of CFRP. *Int. J. Mach. Tools Manuf.* **2008**, *48*, 1464–1473. [CrossRef]

6.   Jiang, Y.; Chen, G.; Zhou, C.; Zhang, Y. Research of carbon fiber reinforced plastic cut by picosecond laser. *Laser Technol.* **2017**, *41*, 821–825.

7.   Ye, Y.; Jia, S.; Xu, Z. Research on Hole Drilling in Carbon Fiber Reinforced Composite by Using Laser Cutting Method. *Aviat. Manuf. Technol.* **2019**, *62*, 50–55.

8.   Song, S. Experimental study on laser cutting of carbon fiber composites. *Mech. Manuf.* **2015**, *53*, 49–51.

9. Hua, Y.; Xiao, T.; Xue, Q. Experimental study about laser cutting of carbon fiber reinforced polymer. *Laser Technol.* **2013**, *37*, 565–570.
10. Zhang, X.; Shen, J.; Wang, J. Study on the process of picosecond laser machining CFRP. *Appled Laser* **2020**, *40*, 86–90.
11. Bluemel, S.; Jaeschke, P.; Suttmann, O. Comparative Study of Achievable Quality Cutting Carbon Fibre Reinforced Thermoplastics Using Continuous Wave and Pulsed Laser Sources. *Phys. Procedia* **2014**, *56*, 1143–1152. [CrossRef]
12. Negarestani, R.; Lin, L.; Sezer, H. Nano-second pulsed DPSS Nd:YAG laser cutting of CFRP composites with mixed reactive and inert gases. *Int. J. Adv. Manuf. Technol.* **2010**, *49*, 553–566. [CrossRef]
13. Li, M.; Gan, G.; Zhang, Y. Thermal defect characterization and strain distribution of CFRP laminate with open hole following fiber laser cutting process. *Opt. Laser Technol.* **2020**, *122*, 105891. [CrossRef]
14. Lau, W.; Lee, W.; Pang, S. Pulsed Nd: YAG Laser Cutting of Carbon Fibre Composite Materials. *CIRP Ann. Manuf. Technol.* **1990**, *39*, 179–182. [CrossRef]
15. Mathew, J.; Goswami, G.; Ramakrishnan, N. Parametric studies on pulsed Nd:YAG laser cutting of carbon fibre reinforced plastic composites. *J. Mater. Process. Technol.* **1999**, *89–90*, 198–203. [CrossRef]
16. Ouyang, W.; Jiao, J.; Xu, Z. Experimental study on CFRP drilling with the picosecond laser "double rotation" cutting technique. *Opt. Laser Technol.* **2021**, *142*, 107238. [CrossRef]
17. Zou, Q.; Jiao, J.; Xu, J.; Sheng, L.; Zhang, Y.; Ouyang, W.; Xu, Z.; Zhang, M.; Xia, H.; Tian, R.; et al. Effects of laser hybrid interfacial pretreatment on enhancing the carbon fiber reinforced thermosetting composites and TC4 alloy heterogeneous joint. *Mater. Today Commun.* **2022**, *30*, 103142.

Article

Study on the Grooved Morphology of CMC-SiC$_f$/SiC by Dual-Beam Coupling Nanosecond Laser

Tao Chen [1,2], Xiaoxiao Chen [2,3,*], Xuanhua Zhang [1,2], Huihui Zhang [2,4], Wenwu Zhang [2,3] and Ganhua Liu [1,*]

1   School of Mechanical and Electrical Engineering, Jiangxi University of Science and Technology, Ganzhou 341000, China
2   Ningbo Institute of Materials Technology and Engineering, Chinese Academy of Sciences, Zhejiang Provincial Key Laboratory of Aero Engine Extreme Manufacturing Technology, Ningbo 315201, China
3   University of Chinese Academy of Sciences, Beijing 100049, China
4   School of Mechanical Engineering and Mechanics, Ningbo University, Ningbo 315211, China
*   Correspondence: chenxiaoxiao@nimte.ac.cn (X.C.); liuganhua1973@163.com (G.L.)

Abstract: Due to the excellent properties of high hardness, oxidation resistance, and high-temperature resistance, silicon carbide fiber reinforced silicon carbide ceramic matrix composite (CMC-SiC$_f$/SiC) is a typical difficult-to-process material. In this paper, according to the relationship between the spatial posture of dual beams and the direction of the machining path, two kinds of scanning methods were set up. The CMC-SiC$_f$/SiC grooving experiments were carried out along different feeding directions (transverse scanning and longitudinal scanning) by using a novel dual-beam coupling nanosecond laser, and the characteristics of grooving morphology were observed by Laser Confocal Microscope, Scanning Electron Microscopy (SEM), and Energy Dispersive Spectrometer (EDS). The results show that the transverse scanning grooving section morphology is V shape, and the longitudinal scanning groove section morphology is W shape. The grooving surface depth and width of transverse scanning are larger and smaller than that of longitudinal scanning when the laser parameters are the same. The depth of the transverse grooving is greater than that of the longitudinal grooving when the laser beam is transverse and longitudinal scanning, the maximum grooving depth is approximately 145.39 µm when the laser energy density is 76.73 J/cm^2, and the minimum grooving depth is approximately 83.76 µm when the laser energy density is about 29.59 J/cm^2. The thermal conductivity of fiber has a significant effect on the local characteristics of the grooved morphology when using a medium energy density grooving. The obvious recasting layer is produced after the laser is applied to CMC-SiC$_f$/SiC when using a high energy density laser grooving, which directly affects the grooved morphology.

Keywords: ceramic matrix composite; dual-beam coupling; nanosecond laser; grooved morphology; high-temperature oxidation

Citation: Chen, T.; Chen, X.; Zhang, X.; Zhang, H.; Zhang, W.; Liu, G. Study on the Grooved Morphology of CMC-SiC$_f$/SiC by Dual-Beam Coupling Nanosecond Laser. *Materials* **2022**, *15*, 6630. https://doi.org/10.3390/ma15196630

Academic Editor: Andres Sotelo

Received: 2 September 2022
Accepted: 20 September 2022
Published: 24 September 2022

Publisher's Note: MDPI stays neutral with regard to jurisdictional claims in published maps and institutional affiliations.

1. Introduction

Silicon carbide fiber-reinforced silicon carbide ceramic matrix composite (CMC-SiC$_f$/SiC) has excellent mechanical properties and stable chemical properties [1]. It has a wide application prospect in the fields of hypersonic aircraft [2], the nuclear energy industry [3], and the national defense industry [4].

Because CMC-SiC$_f$/SiC is a difficult-to-process material [5] and laser processing has the advantages of non-contact, greening, and high precision, domestic and foreign scholars have gradually paid attention to laser processing CMC-SiC$_f$/SiC. Zhai et al. [6] used a femtosecond laser to etch carbon fiber reinforced silicon carbide ceramic matrix composites (CMC-C$_f$/SiC) and found that the laser removal mechanism could be changed from photothermal effect to photochemical effect by changing the laser parameters. Moreover, Zhai et al. [7,8] found through further research that the femtosecond laser refracts and scatters on the surface of the material when the surface of the material is uneven, resulting in the narrowing or bending of the microgrooves. Nasiri et al. [9] think that the oxidation process of CMC-SiC$_f$/SiC was related to temperature and time. The differences in the structure of fiber composites and the properties of each component affect the laser removal mechanism of fiber composites and the surface morphology [10]. Zhang et al. [11] used a millisecond laser to ablate CMC-SiC$_f$/SiC and found that the ablation area has problems such as fiber fracture, interface–matrix debonding, interface–fiber debonding, and fiber hollowing. The evolution mechanism is closely related to the internal stress caused by the temperature gradient in the ablation region and the melting temperature difference of the material components. Wei et al. [12] studied the removal mechanism of CMC-SiC$_f$/SiC by underwater femtosecond laser ablation and found that the removal mechanism of CMC-SiC$_f$/SiC was the decomposition of SiC; they believe that the SiC matrix was first removed, and the interface layer was then removed, and the SiC fibers were finally removed. Du et al. [13] ablated CMC-SiC$_f$/SiC with oxygen-acetylene flame and found that the material removal mechanism is dominated by the sublimation of the matrix and fibers, and the ablation process is accompanied by mechanical spalling and material oxidation. Liu et al. [14] used millisecond laser processing of CMC-SiC$_f$/SiC and found that three kinds of tiny particles with different diameters were attached to the surface of the recasting layer, spherical micro protrusion (20–48 μm), bubble particle (5–15 μm), and submicron particles (<1 μm), respectively. Jiao et al. [15] used a nanosecond laser to remove CMC-C$_f$/SiC and found that heat accumulation was the main factor in removing CMC-C$_f$/SiC; in addition, the heat was mainly transferred along the fiber arrangement direction.

In addition to studying the mechanism of laser removal of ceramic matrix composite (CMC), scholars also analyzed the characteristics of different types of laser-processed fiber composite materials. Gavalda Diaz et al. [16] suggested that various types of lasers to remove CMC would generate heat affected zones (HAZ), but the width of the HAZ could be effectively reduced by using a laser with shorter pulse width. Takahashi et al. [17] used Infrared (IR) and Ultraviolet (UV) lasers to groove carbon fiber reinforced plastic (CFRP) and found that UV laser irradiation produced a narrower HAZ with better processing quality compared to IR lasers. Therefore, the width of the HAZ can be reduced, and higher processing quality can be obtained by selecting the laser parameters [18].

Domestic and foreign scholars have reported on beam-coupling laser removal materials. Chen et al. [19] proposed a new method for machining stainless steel using using the dual-beam coupling nanosecond laser. Zhou et al. [20] coupled continuous laser and pulsed laser at one point to polish S136H die steel with a free-form surface and obtained lower surface roughness. Jiang et al. [21] used a three-beam coupling laser for cladding experiments and achieved better results. Liu et al. [22] established a mathematical model of the relationship between the three-beam coupling laser cladding process parameters and the surface morphology of the cladding layer and obtained a cladding layer with better surface morphology. Therefore, coupling laser processing has unique advantages.

In this paper, CMC-SiC$_f$/SiC grooves were carried out along different feed directions by using a new dual-beam coupling nanosecond laser in an air environment. The mechanism of CMC-SiC$_f$/SiC dual-beam coupling nanosecond laser grooving was analyzed based on the study of the change in the morphology characteristics of laser grooving. This work expanded the process types of CMC-SiC$_f$/SiC laser processing, which can provide technical support for high-performance applications of CMC.

2. Materials and Methods

The CMC-SiC$_f$/SiC workpiece (100 mm × 20 mm × 2 mm) was used for the dual-beam coupling nanosecond laser grooving experiments. The CMC-SiC$_f$/SiC workpiece is composed of SiC fiber, SiC matrix, and interface layer; SiC fiber is arranged in vertical and horizontal weaving manner to form a fiber preform, the interface layer is wrapped on the surface of the SiC fiber, and the CMC-SiC$_f$/SiC workpiece is prepared by chemical vapor infiltration (CVI) process, as shown in Figure 1. The diameter of the SiC fiber is about 12–13 μm, and the thickness of the interface layer is approximately 0.4 μm. The preparation process of CMC-SiC$_f$/SiC mainly includes the CVI process, precursor impregnation and pyrolysis (PIP) process, and melt infiltration (MI) process. The advantage of the CVI process is that there is basically no damage to the fibers, while the disadvantage is that the prepared composites have high porosity.

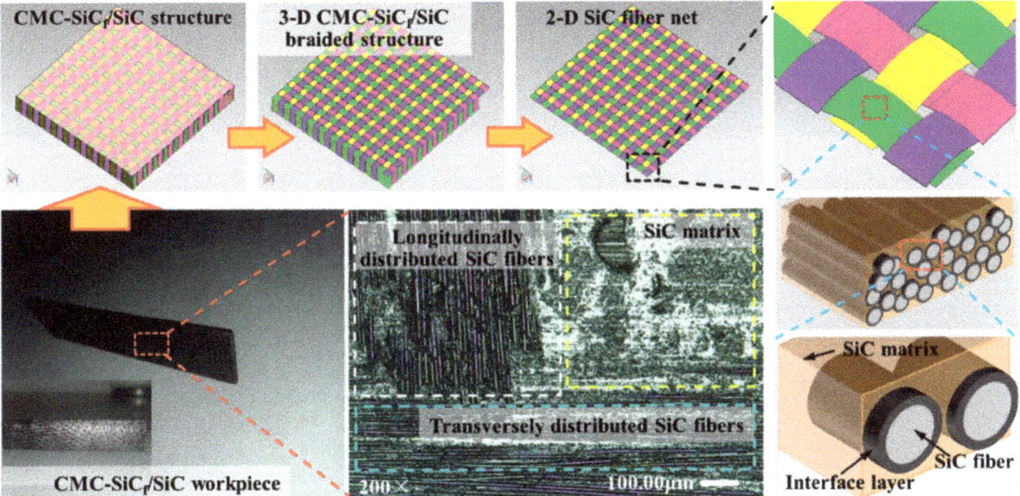

Figure 1. The structure and the surface morphology of CMC-SiC$_f$/SiC.

The schematic diagram of CMC-SiC$_f$/SiC is processed by a nanosecond laser, as shown in Figure 2. Processing equipment mainly includes a four-axis motion platform (Model: D-200CNC, China), nanosecond laser, optical platform, optical path system components, control system, etc., as shown in Figure 2a. The pulse width (τ) is 78 ns, the wavelength (l) is 532 nm, and the beam quality (M^2) less than or equal to 1.5. M^2; is used to describe the quality of the laser beam. The ideal beam quality M^2; is 1, it can be greater than or equal to 1 in practice, and the beam quality of a laser source becomes worse as M^2 increases.

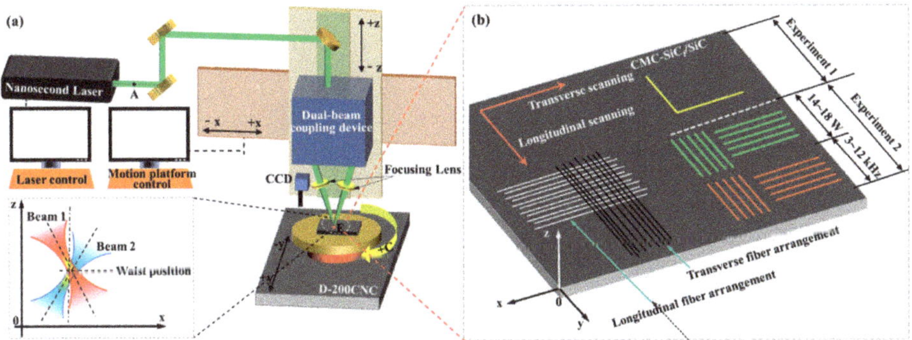

Figure 2. Grooving experimental platform and grooving experimental parameters. (**a**) Laser processing platform; (**b**) grooving experimental parameters.

The spot quality analyzer was placed at point "A" to analyze the beam quality of the laser source, and the laser beam energy density distribution satisfied the Gaussian distribution, as shown in Figure 3.

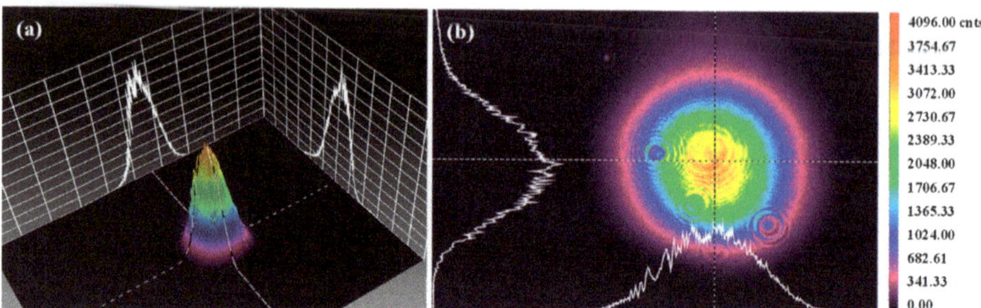

Figure 3. Energy density distribution in the spot of the experimental laser source. (**a**) Three-dimensional energy density distribution; (**b**) two-dimensional energy density distribution.

The single beam was first split into two laser beams, and finally, the two beams were coupled at point "E" by adjusting the angle of the two laser beams. The angles of the two laser beams of the dual-beam coupling nanosecond laser in the y-0-z plane were 1.09 degrees and 1.33 degrees, respectively, and the angle difference between the two beams was 0.24 degrees. The angles in the x-0-z plane are 3.58 degrees and 9.97 degrees, respectively, and the sum of the angles between the two laser beams is 13.55 degrees. Because the focal lengths of the two optical focusing lenses used in the experiment are equal, the two laser beams have the same focal position; that is, the two beams have the same spot size in the beam waist plane of the single beam. However, since the incidence angles of the beams in the x-0-z plane are not equal, the coupled poses of the two laser beams in the x-0-z plane are shown in Figure 2. When the laser pulse energy emitted by the laser source is constant, the pulse energy of the two laser beams is not equal, and the pulse energy of the right laser beam is smaller than the left one.

Fiber composites have high thermal anisotropy in the direction parallel to and perpendicular to the fiber [23]. In the plane, the heat transfer is obvious along the direction parallel to the fiber arrangement, and in space, the heat can be transmitted in multiple directions, including the direction perpendicular to the fiber arrangement [24]. While the substrate absorbs heat, it also transfers heat to the interface layer, which in turn transfers heat to the fiber. In addition to the heat absorbed by the fiber itself from the laser, the fiber also conducts heat from the interface layer. Due to the thermal conductivity of fibers, heat is more easily conducted along the direction of fiber distribution [17].

In order to distinguish the intensity of laser energy density during the double-beam coupling nanosecond laser grooving, the grooving energy region is divided into low energy density grooving, medium energy density grooving, and high energy density grooving. In a certain range of laser energy density, when the grooving track is discontinuous, it is called low energy density grooving, and this range can be called the low energy density region. When the grooving track is continuous, but the recasting layer cannot completely cover the grooving surface, it is called medium energy density grooving, and this interval can be called medium energy density region. When the grooving track is continuous, and the recasting layer completely covers the grooving surface, it is called high energy density grooving, and this interval can be called high energy density region.

Two groups of grooving experiments were carried out in two different regions of the surface of the CMC-SiC$_f$/SiC sample: experiments 1 and 2, as shown in Figure 2b. The laser power was measured using the laser power meter at point "E" in Figure 2a. In order to explore the influence of fiber thermal conductivity on grooved morphology, grooving experiment 1 was carried out, as shown in Table 1. In order to explore the influence of different scanning directions of dual-beam coupling nanosecond laser on grooved morphology, experiment 2 was carried out on the basis of grooving experiment 1, as shown in Table 2. In experiment 2, the purpose of changing the laser energy density was achieved by changing the laser power when the laser repetition frequency was 10 kHz, and the laser repetition frequency was changed to achieve the purpose of changing the laser energy density when the laser power was 14 W. The feed speeding (v) was 200 mm/min, and the defocus amount was 0 mm in experiments 1 and 2.

Table 1. Experiment parameters of nanosecond laser grooving (experiment 1, medium energy density grooving).

Parameter	Numerical Value	
Laser power, P/W	10	
Repetition frequency, f/kHz	20	
laser energy density, φ/(J/cm^2)	8.22	
Scanning direction	Longitudinal	Transverse

Table 2. Experiment parameters of nanosecond laser grooving (experiment 2, high energy density grooving).

Parameter	Numerical Value				
Laser power, P/W	14	15	16	17	18
Repetition frequency, f/kHz	3	5	7	10	12
Scanning direction		Longitudinal		Transverse	

The laser energy density is expressed as [25,26]:

$$\varphi = \frac{P}{\pi f \omega_0^2} \tag{1}$$

where P is the average laser power (W), f is the laser repetition frequency (Hz), and ω_0 is the equivalent waist radius of 44 μm for a dual-beam coupling laser. According to Equation (1), the laser energy densities corresponding to different laser repetition frequencies and laser powers can be calculated as shown in Tables 3 and 4, respectively.

Table 3. The laser energy density corresponding to different laser powers when the repetition frequency is 10 kHz.

Laser Power, P/W	14	15	16	17	18
laser energy density, φ/(J/cm^2)	23.01	24.66	26.31	27.95	29.59

Table 4. The laser energy density corresponding to different laser repetition frequencies when the laser power is 14 W.

Repetition Frequency, f/kHz	3	5	7	10	12
laser energy density, φ/(J/cm^2)	76.73	46.04	32.88	23.01	19.18

After the grooving experiments of the two groups, the specimens were immersed in a solution of 99.7% ethanol for 10 min for ultrasonic cleaning, and then the grooved morphology was observed by Laser Confocal Microscope (Model: KEYENCE VK-X200K, Japan), Scanning Electron Microscopy (SEM, Model: QUANTA FEG 250, FEI, Hillsboro, OR, USA 250), and Energy Dispersive Spectrometer (EDS, Model: FEI, Hillsboro, OR, USA).

3. Results and Discussions

3.1. Analysis of Grooved Topographies of Medium Energy Density Grooving

In order to analyze the physical and chemical changes in the surface grooving regions, the grooved morphologies of experiment 1 were observed under laser confocal microscopy, as shown in Figure 4. Under the parameters of experiment 1 ($\varphi = 8.22$ J/cm^2), grooves in different scanning directions are continuous. When the laser energy density is 8.22 J/cm^2, CMC-SiC$_f$/SiC can be effectively removed, but the recast layer cannot completely cover the groove surface, which is the grooves with medium energy density. When the nanosecond laser acts on the surface of the material, the high temperature triggers a series of chemical reactions, such as the formation of oxides, and finally, leads to the removal of CMC-SiCf/SiC under the influence of various physical and chemical factors. In the process of nanosecond laser grooving, there is an oxidation phenomenon, and the white solid particles produced are SiO$_2$ [27].

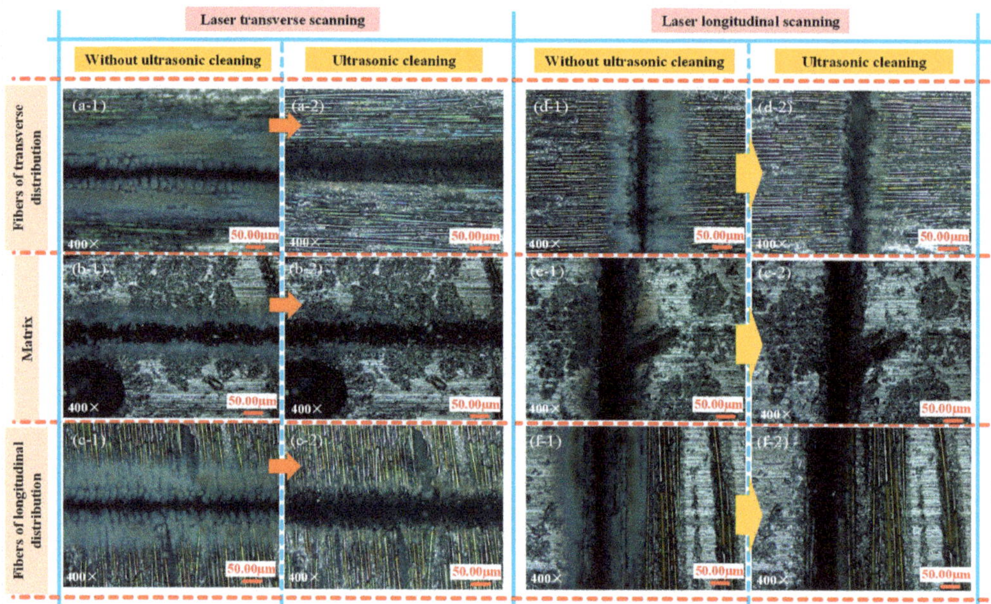

Figure 4. The grooved morphologies of different regions in experiment 1 (**a-1**–**f-2**).

It was found that the orientation of the fibers could affect the grooved morphology under confocal microscopy, as shown in Figure 5, as it correlates with the thermal conduction velocity of the fibers being higher than that of the matrix and a large difference in the removal temperature of each component of the composite [12,13]. After the light beam is irradiated to the fiber, the fiber can conduct heat along the fiber arrangement direction so that the groove surface morphology of the fiber arrangement area is different from that of the matrix area. Therefore, the orientation and thermal conductivity of the fibers must be considered when processing CMC-SiC$_f$/SiC by nanosecond laser at medium energy density.

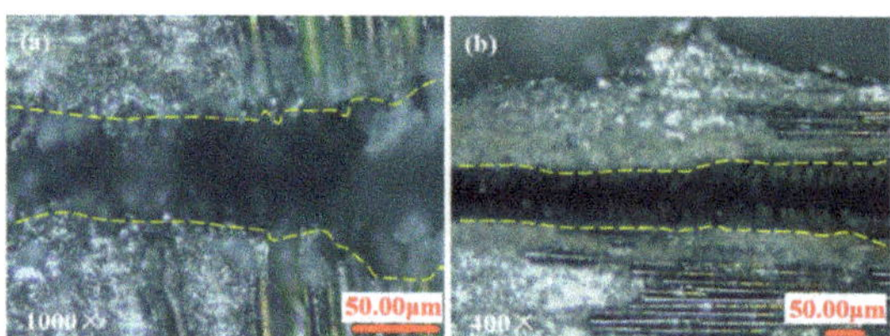

Figure 5. The morphologies of transverse scanning grooves in different regions. (**a**) Regions of matrix and longitudinal fiber arrangement; (**b**) regions of matrix and transverse fiber arrangement.

3.2. Analysis of Three-Dimensional Feature of High Energy Density Grooving

3.2.1. Three-Dimensional Feature of Grooves at Different Laser Energy Densities

The laser energy density can be increased by increasing laser power and decreasing repetition frequency. It was found that when laser power is 14 W and the repetition frequency is 12 kHz, the energy density φ is 19.18 J/cm^2, and a large number of recasting layers are accumulated on both sides of the grooving surface. CMC-SiC$_f$/SiC removed by fiber heat conduction is completely covered by the recasting layer (See Figure 6), which is the high energy density grooving. The depth of the groove bottom on the left is significantly larger than that on the right when the laser is longitudinal scanning, which is caused by the unequal pulse energy of the two laser beams.

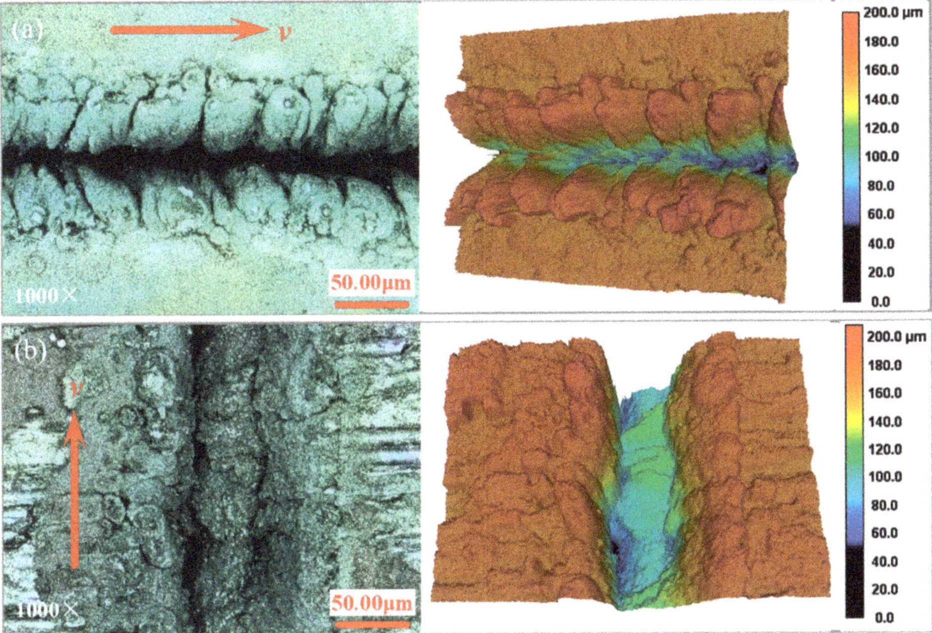

Figure 6. The grooved morphologies in different scanning directions under laser power of 14 W and repetition frequency of 12 kHz. (**a**) Transverse scanning region; (**b**) longitudinal scanning region.

The values of grooving width of longitudinal scanning (GWLS), grooving width of transverse scanning (GWTS), grooving depth of longitudinal scanning (GDLS), and grooving depth of transverse scanning (GDTS) in the fiber longitudinal distribution region (FLDR), fiber transverse distribution region (FTDR), and matrix region (MR) are shown in Figure 7. It was found that the trough depth and width in each region are close to the overall value, and the error of each set of parameters is small. This is because the recast layer covers the grooved surface, and the width and depth of the groove are mainly affected by the morphology of the recasting layer, while the second factor is the thermal conductivity of the fiber. That is, when the energy density φ is greater than or equal to 19.18 J/cm^2, the morphology of CMC-SiC$_f$/SiC dual-beam coupling nanosecond laser etching is less affected by the thermal conductivity of fibers. The recasting layer produced by laser on CMC-SiC$_f$/SiC is obvious, which directly affects the grooved morphologies.

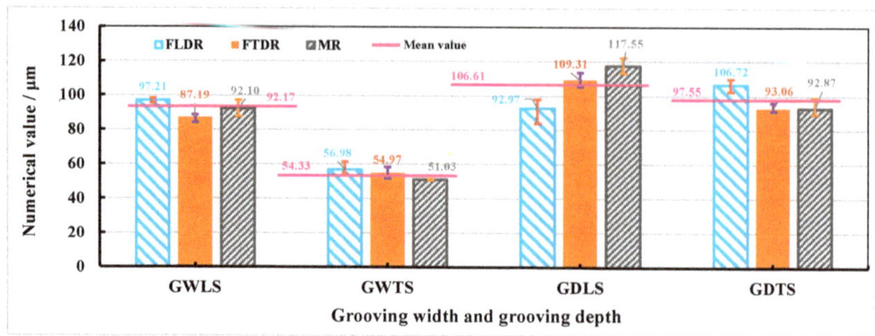

Figure 7. Grooving depth and width in different regions under laser power of 14 W and repetition frequency of 12 kHz.

Figure 8 shows the grooved section morphologies under different frequencies and scanning directions when laser power is 14 W and defocus amount is 0 mm. It can be seen that the groove shape in transverse scanning is a V shape and that in longitudinal scanning is a W shape. Compared with longitudinal scan grooving, the depth (h) and width (s) of transverse scan grooving are larger and smaller, respectively. This is because the trajectory of the two laser beams incident from the cross-section of the grooving and acting on the bottom of the grooving almost coincide when the laser beam is transverse scanning. Therefore, a secondary grooving effect in space exists. Two laser beams incident from two directions of the cross-section of the groove and act on both sides of the groove simultaneously when the laser beam is longitudinal scanning. Therefore, there is no secondary grooving effect in space.

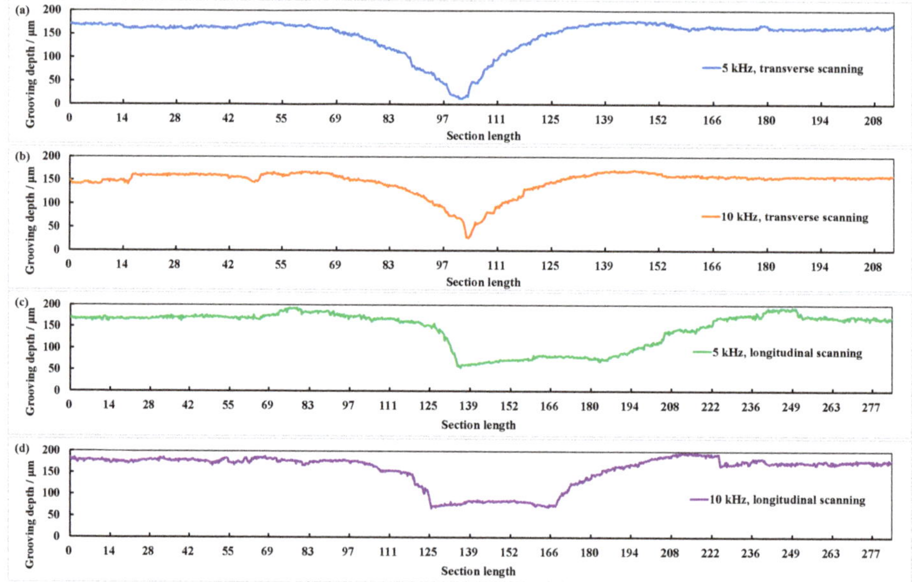

Figure 8. The section morphologies of grooves in different scanning directions. (**a**) Transverse scanning and repetition frequency of 5 kHz; (**b**) transverse scanning and repetition frequency of 10 kHz; (**c**) longitudinal scanning and repetition frequency of 5 kHz; (**d**) longitudinal scanning and repetition frequency of 10 kHz.

When the laser beam is transverse scanning, the grooving depth and grooving width show a trend of first increasing, then decreasing, and finally increasing with the increase in laser energy density. When the beam is longitudinal scanning, the grooving depth first decreases and then increases with the increase in laser energy density, while the grooving width first increases, then decreases, and finally increasing with the increase in laser energy density, as shown in Figure 9. The depth of the transverse grooving is greater than that of the longitudinal grooving when the laser beam is transverse and longitudinal scanning, the maximum grooving depth is approximately 145.39 µm when the laser energy density is 76.73 J/cm^2, and the minimum grooving depth is approximately 83.76 µm when the laser energy density is about 29.59 J/cm^2 (See Figure 9a). The width of the transverse grooving is smaller than that of the longitudinal grooving when the laser beam is transverse and longitudinal scanning; the maximum grooving width is approximately 132.40 µm when the laser energy density is about 29.59 J/cm^2, the minimum grooving width is approximately 54.33 µm and when the laser energy density is about 19.18 J/cm^2 (See Figure 9b).

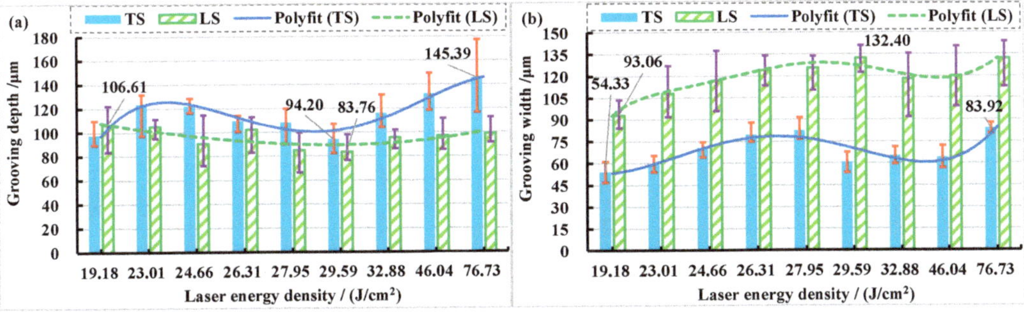

Figure 9. The changes in grooving depth and width in different laser energy densities. (**a**) Grooving depth; (**b**) grooving width.

3.2.2. Influence of Different Laser Powers on the Grooved Morphologies

The grooved morphology is shown in Figure 10 when the repetition frequency is 10 kHz, the laser power is 14 W, the feed speed is 200 mm/min, and transverse scanning. Recasting layers with irregular morphology are generated on both sides of the groove when the energy density is high; this is 23.01 J/cm^2. This is because when the laser energy density is high enough, CMC-SiCf/SiC melts and is removed. As CMC-SiCf/SiC cools and solidifies in a very short time after the light beam leaves the material irradiation region, it rapidly shrinks irregularly from a molten state to a solid state, forming a recast layer with irregular morphology. Compared with the middle energy density grooves in Figure 5, the recasting layer of high energy density grooves not only covers the grooves completely but also covers some fibers on the material surface. The morphology changes caused by the thermal conductivity of fibers cannot be observed under the confocal microscope at 200×, 400×, and 1000× magnifications. Therefore, it is further indicated that when the laser energy density is high, the obvious recasting layer is produced after the laser is applied to CMC-SiCf/SiC, which directly affects the grooved morphology.

When the laser power is 14 W and the repetition frequency is 10 kHz, that is, the energy density in the spot is 23.01 J/cm^2, the grooved morphologies of laser transverse scanning processing in different regions are shown in Figure 11. The recasting layer completely covers both sides of the grooved surface and presents a "zig-zag" structure along the laser scanning direction, and there are "recessed ravines" on the surface of the recasting layer. The material is in a molten state and is sprayed on both sides of the grooved surface under the condition of high temperature and high pressure. As the temperature and pressure decrease in the laser grooved region, the molten material cools rapidly to form a recasting layer. Under the action of high pressure, the molten material near the laser irradiation sputters around the laser irradiation point and forms a "gully" after cooling. The cooled products are stacked one after another to form a recasting layer. The recasting layer of the grooved surface in the fiber transverse arrangement region and the fiber longitudinal arrangement region presents a flocculent structure (See Figure 11(a-2,b-2)) and is in a fluffy state (See Figure 11(a-4,b-4)). The recasting layer in the matrix distribution region presents a cohesive structure (See Figure 11(c-2)), which is relatively flat compared to the recasting layer in the fiber arrangement region. A small amount of material is removed from the grooved border of the matrix distribution area.

Figure 10. The grooved morphologies with repetition frequency of 10 kHz and laser power of 14 W: (**a**) 200× magnifications; (**b**) 400× magnifications; (**c**) 1000× magnifications.

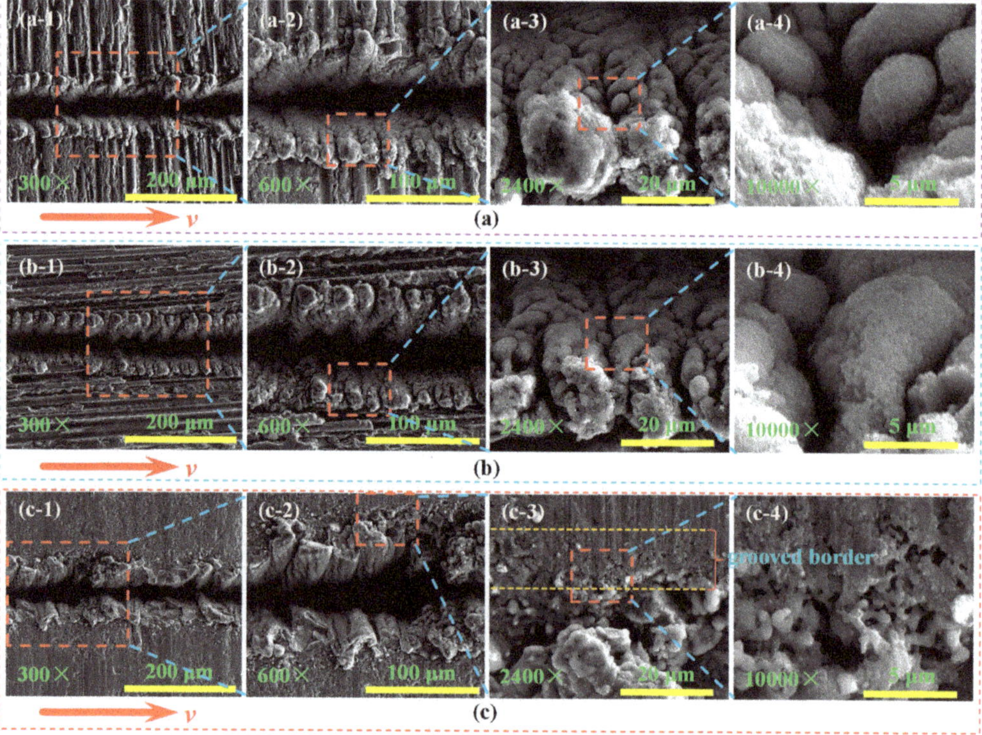

Figure 11. The grooved morphologies in different regions during transverse grooving. (**a-1–a-4**) Fiber longitudinal arrangement; (**b-1–b-4**) fiber transverse arrangement; (**c-1–c-4**) matrix distribution.

The incidence angle of the two laser beams [28] and the angle between the coupling of the two laser beams can affect the element distribution of the recasting layer. Figure 12 shows the grooved morphology when the longitudinal fiber arrangement and laser transverse scanning region. The distribution of Si, O, N, C, and B element on the surface of the recasting layer on the upper side of the grooved surface is denser than that on the lower side. This is because the laser pulse energy irradiated on the material surface by the laser beam with an oblique angle can be divided into three regions during grooving [25]. The laser energy density on the side with the smaller angle between the laser beam and the material surface is higher, resulting in more oxide distribution of the recasting layer on the upper side of the grooved surface.

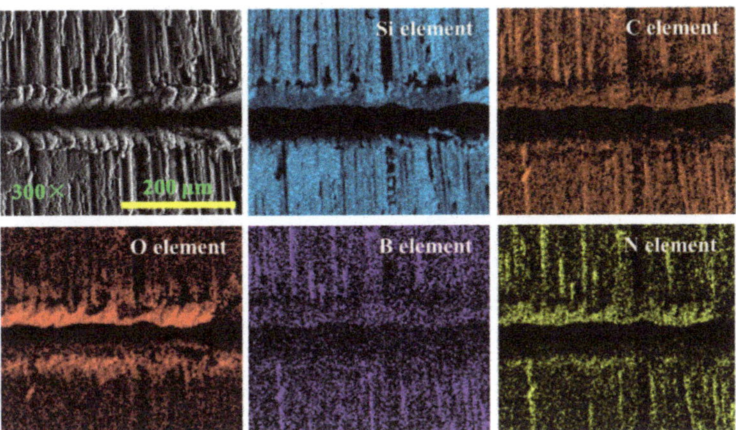

Figure 12. Distribution of different elements of fiber longitudinal arrangement and laser transverse scanning region.

Figure 13 shows the grooved morphology when the fibers are transverse arrangement, and the laser beam is longitudinal scanning. There is a bulge in the center of the grooving, and the recasting layer in the depression in the center of the grooving has cracks under the action of surface tension and residual stress. Because the pulse energy of the two laser beams and the incidence angle in the x-0-z plane are not equal, the distribution of Si, C, O and N elements in the grooved region 1 is denser than that of the grooved region 2. Therefore, the grooved morphology can be changed by changing the pulse energy of the two laser beams and the incidence angle of the laser beams.

The grooved morphologies corresponding to different laser powers are shown in Figure 14 when the longitudinal fiber arrangement and laser transverse scanning. The "serrated" tooth profile of the recasting layer becomes more pronounced as the laser power increases. The two laser beams of the dual-beam coupling nanosecond laser are irradiated obliquely on the material surface in the y-0-z plane, and the recasting layer on the upper side of the grooved surface has a more obvious tooth shape.

The relationship between grooving depth and grooving width under different laser powers is shown in Figure 15 when the repetition frequency is 10 kHz. It was found that compared with longitudinal scanning (LS), transverse scanning (TS) has a larger depth and smaller width. The grooving depths of the transverse and longitudinal scanning of the laser beam first increased and then decreased with the increase in laser power. This is because the plasma gradually increases with increasing laser power, resulting in a plasma shielding effect [29]. The plasma shielding effect hinders the further action of the laser and CMC-SiC$_f$/SiC. However, the amount of recasting layer and oxide generated increases with laser power, and these formations adhere to the grooved surface, ultimately leading to a decrease in groove depth. The grooving width of the transverse and longitudinal scanning of the laser beam first increased and then decreased with the increase in laser power, while the groove width of the longitudinal scanning of the laser beam increased with the increase in the laser power. It can be seen that when the laser beam is transverse scanning, the thickness of the recasting layer covering the grooved surface increases with the increase in laser power, and the thickness of the recasting layer is the largest when the laser power is 18 W, resulting in the smallest grooving width.

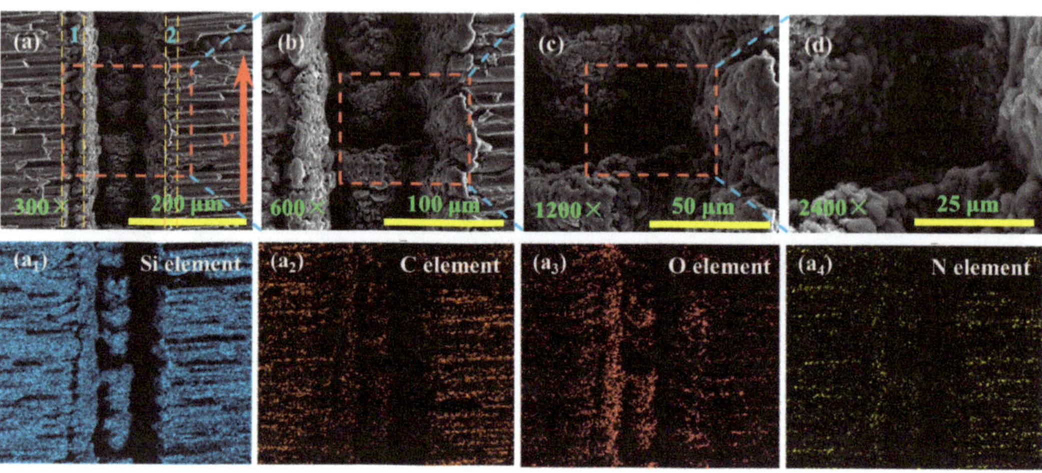

Figure 13. The grooved morphologies corresponding to different magnifications and distribution of different elements of fiber transverse arrangement (a_1–a_4) and laser longitudinal scanning region: (**a**) 300× magnifications; (**b**) 600× magnifications; (**c**) 1200× magnifications; (**d**) 2400× magnifications.

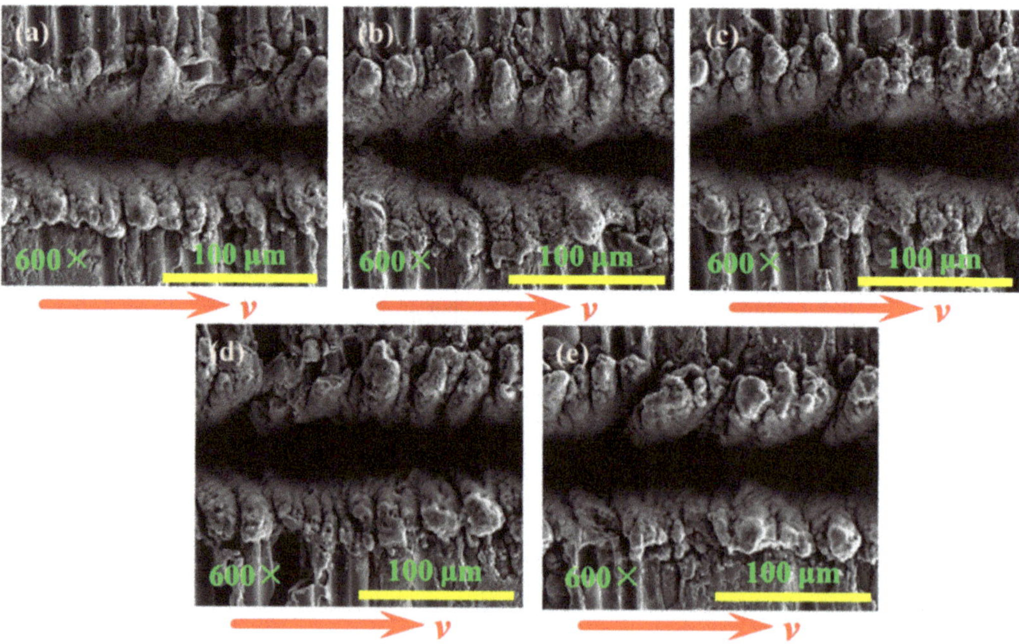

Figure 14. The grooved morphologies corresponding to different laser powers: (**a**) 14 W; (**b**) 15 W; (**c**) 16 W; (**d**) 17 W; (**e**) 18 W.

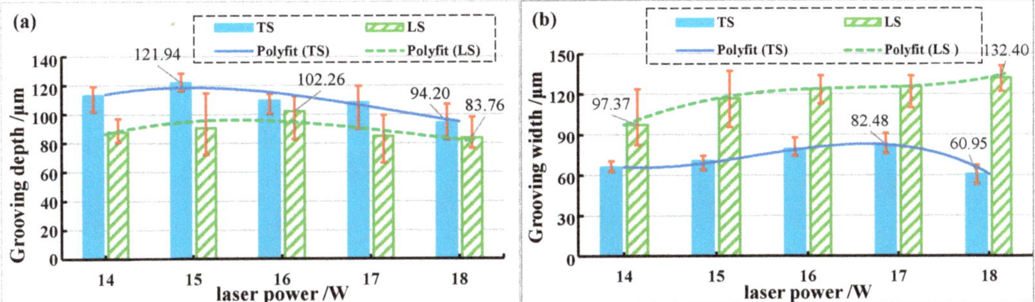

Figure 15. The change in grooving depth and width with increasing laser power in different feeding directions. (**a**) Grooving depth; (**b**) grooving width.

3.2.3. Influence of Different Laser Frequencies on the Grooved Morphologies

The grooved morphologies corresponding to different laser repetition frequencies are shown in Figure 16 when the fiber longitudinal arrangement and laser transverse scanning. The recasting layer of the grooved surface is relatively flat when the repetition frequency is less than 10 kHz. The recasting layer of the grooved surface is more uneven, and the tooth shape is obvious when the repetition frequency is greater than or equal to 10 kHz. The recasting layer of the grooved surface is irregular, and a part of the fibers is not removed when the repetition frequency is 12 kHz.

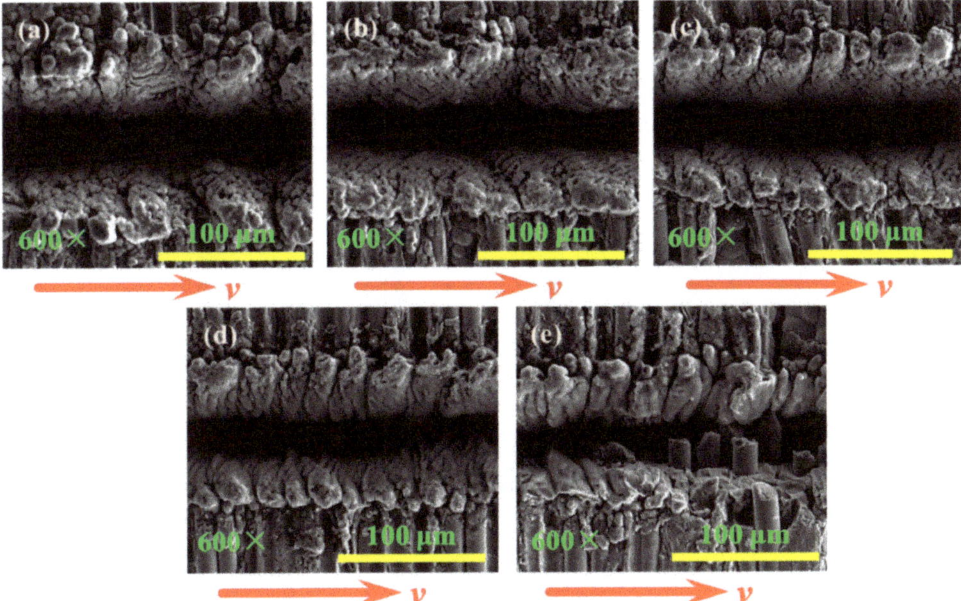

Figure 16. The grooved morphologies corresponding to different repetition frequencies: (**a**) 3 kHz; (**b**) 5 kHz; (**c**) 7 kHz; (**d**) 10 kHz; (**e**) 12 kHz.

The fibers are cut by the laser beam from the side of the grooved surface, as shown in Figure 17, when the repetition frequency is 12 kHz, laser power is 10 W, fibers are longitudinal arrangement, and laser beam is transverse scanning (φ = 19.18 J/cm^2). This is because the fiber can be removed when the energy density in the laser spot is small, and the laser beam is directly irradiated on fiber, but the reflected laser cannot remove the fiber. Compared to SiC matrix, SiC fibers are more difficult to be removed by laser.

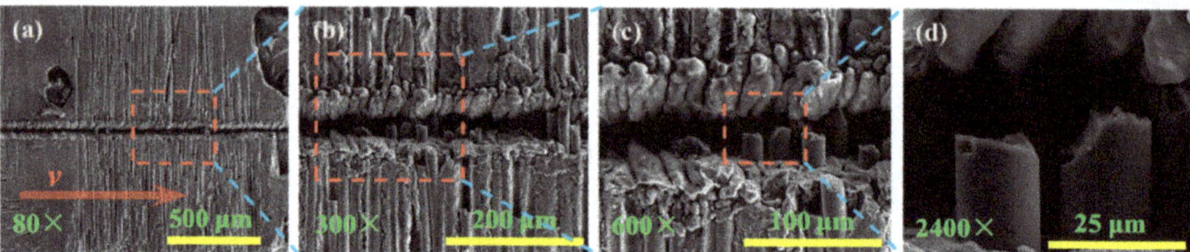

Figure 17. The grooved morphologies corresponding to different magnifications: (**a**) 80× magnifications; (**b**) 300× magnifications; (**c**) 600× magnifications; (**d**) 200× magnifications.

The relationship between grooving depth and grooving width at different repetition frequencies is shown in Figure 18 when the laser power is 14 W. It was found that transverse scanning (TS) has greater depth and narrower width than longitudinal scanning (LS). When the frequency is 3 kHz, the measurement deviation of transverse scanning depth is large. This is because the repetition frequency is smaller so that the energy density in the unit area is more concentrated, the number of pulses at the same time is smaller, and the spacing and depth of pulse points at the bottom of the tank are larger. The fluctuation range of measured values increases as the degree of unevenness increases (See Figure 5a). Longitudinal scanning of the grooved morphology shows that the depth on both sides of the bottom of the grooving is not equal (See Figure 5b). This is because when the beam irradiates to the bottom of the grooving, it is already in a positive defocusing state, and the energy density of the two laser beams from the beam splitter is not equal.

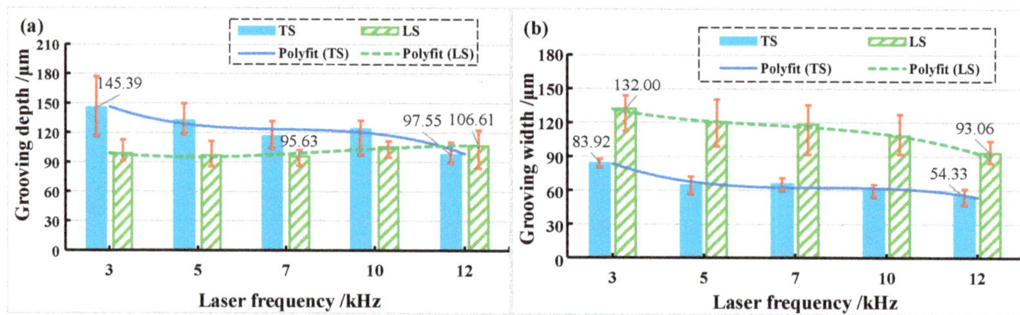

Figure 18. The change in grooving depth and width with increasing frequency in different feeding directions. (**a**) Grooving depth; (**b**) grooving width.

4. Grooving Mechanism

The above analysis can provide theoretical support for the study of the CMC-SiC$_f$/SiC dual-beam coupling nanosecond laser grooving process. In the process of nanosecond laser grooving, the thermal effect plays a major role and forms the laser action region. Heat diffuses from the center of the laser action area to the periphery to form the HAZ [18].

When CMC-SiC$_f$/SiC is melted at a high temperature, the material is oxidized to SiO_2 and aggregates to form solids of granular. Under high temperature and high pressure, SiO_2 particles are continuously sprayed outward from the laser action area and attached to the grooving area under the action of gravity to form the grooving affected area. CO_2, CO, and other gases generated during the grooving process are diffused into the air. Under medium energy density, the grooved morphology is significantly affected by the thermal conductivity of fibers, and the grooved surface morphology is irregular. Under high energy density, there is a difference between transverse scanning and longitudinal scanning in the grooving section morphology. Compared with the longitudinal scanning, the depth *h* of the transverse scanning grooving surface is larger, and the width *s* is smaller. The grooved morphology of transverse scanning is V shape, and that of longitudinal scanning is W shape, as shown in Figure 19.

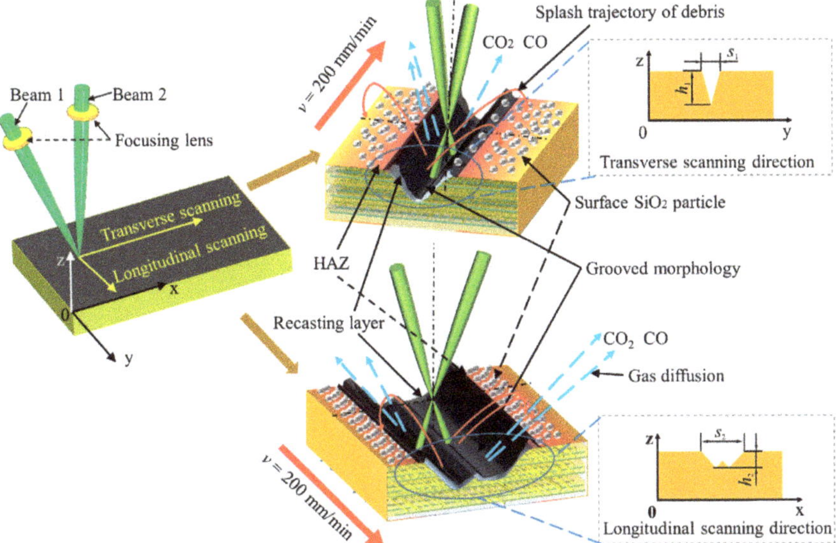

Figure 19. Schematic diagram of CMC-SiC$_f$/SiC by using dual-beam coupling nanosecond laser.

5. Conclusions

In this paper, two groups of grooves of CMC-SiC$_f$/SiC are carried out by using a new dual-beam coupling nanosecond laser. The effects of processing parameters such as scanning direction, laser power, and laser repetition frequency on the grooved morphology were studied, and the material removal mechanism and grooved morphology in different regions of CMC-SiC$_f$/SiC were analyzed. The following conclusions are drawn:

(1) The thermal conductivity of fiber has a significant effect on the local characteristics of the grooved morphology when using a medium energy density grooving, such as the laser energy density is 8.22 J/cm^2. The obvious recasting layer is produced after the laser is applied to CMC-SiC$_f$/SiC when using a high energy density laser grooving, which directly affects the grooved morphology. The thermal conductivity of fiber has a significant effect on the local characteristics of the grooved morphology when using a medium energy density grooving, such as the laser energy density being greater than or equal to 19.18 J/cm^2. When the dual-beam coupling nanosecond laser irradiated to CMC-SiC$_f$/SiC, a small portion of solid CMC-SiC$_f$/SiC was directly converted to gas and sublimated to remove, and most of the solid CMC-SiC$_f$/SiC was removed when heated;

(2) The depth of the transverse grooving is greater than that of the longitudinal grooving when the laser beam is transverse and longitudinal scanning, the maximum grooving depth is approximately 145.39 μm when the laser energy density is 76.73 J/cm^2, and the minimum grooving depth is approximately 83.76 μm when the laser energy density is about 29.59 J/cm^2. The plasma shielding effect affects the grooving depth of dual-beam coupling nanosecond lasers. The fiber can be removed when the energy density in the laser spot is small, and the laser beam is directly irradiated on SiC fiber, but the reflected laser cannot remove the fiber. Compared to SiC matrix, SiC fibers are more difficult to be removed by laser;

(3) The grooved morphology of transverse scanning is V shape, and that of longitudinal scanning is W shape. In the transverse scanning of the dual-beam coupling nanosecond

laser, the rejections of the two beams incident from the cross-section of the groove and acting on the bottom of the grooves almost coincide, and one is in front of the other. This is because of the existence of a secondary grooving effect in time. During the longitudinal scanning, two laser beams of light incident from two directions of the cross-section of the slot act on both sides of the slot and are synchronized simultaneously. This is because there is no secondary grooving effect. Therefore, the grooving section morphology of transverse scanning has a larger depth and smaller width than that of longitudinal scanning when the laser parameters are the same.

Author Contributions: Conceptualization, methodology, and validation, T.C., X.C. and X.Z.; formal analysis, T.C., X.Z. and G.L.; resources, X.C. and W.Z.; writing—original draft preparation, T.C. and X.C.; writing—review and editing, T.C., X.C. and H.Z.; visualization, T.C., X.C. and G.L.; supervision, X.C. and G.L.; project administration, X.C. and W.Z. All authors have read and agreed to the published version of the manuscript.

Funding: This work was financially supported by the Natural Science Foundation of Zhejiang Province (LY20E050004), the Zhejiang Provincial Key Research and Development Program (2020C01036), and the "Science and Technology Innovation 2025" Major Project of Ningbo City (2019B10074).

Institutional Review Board Statement: Not applicable.

Informed Consent Statement: Not applicable.

Data Availability Statement: Not applicable.

Acknowledgments: The authors gratefully acknowledge Ningbo Institute of Materials Technology and Engineering, Chinese Academy of Sciences.

Conflicts of Interest: The authors declare no conflict of interest.

References

1. Zhang, S.; Feng, Y.; Gao, X.; Song, Y.; Wang, F.; Zhang, S. Modeling of fatigue failure for SiC/SiC ceramic matrix composites at elevated temperatures and multi-scale experimental validation. *J. Eur. Ceram. Soc.* **2022**, *42*, 3395–3403. [CrossRef]
2. Ruggles-Wrenn, M.B.; Jones, T.P. Tension–compression fatigue of a SiC/SiC ceramic matrix composite at 1200 °C in air and in steam. *Int. J. Fatigue* **2013**, *47*, 154–160. [CrossRef]
3. Lawal, S.S.; Ademoh, N.A.; Bala, K.C.; Abdulrahman, A.S. A Review of the Compositions, Processing, Materials and Properties of Brake Pad Production. *J. Phys. Conf. Ser.* **2019**, *1378*, 032103. [CrossRef]
4. Liu, Y.; Quan, Y.; Wu, C.; Ye, L.; Zhu, X. Single diamond scribing of SiCf/SiC composite: Force and material removal mechanism study. *Ceram. Int.* **2021**, *47*, 27702–27709. [CrossRef]
5. Li, L. Fatigue Damage and Lifetime of SiC/SiC Ceramic-Matrix Composite under Cyclic Loading at Elevated Temperatures. *Materials* **2017**, *10*, 371. [CrossRef]
6. Zhai, Z.; Wei, C.; Zhang, Y.; Cui, Y.; Zeng, Q. Investigations on the oxidation phenomenon of SiC/SiC fabricated by high repetition frequency femtosecond laser. *Appl. Surf. Sci.* **2020**, *502*, 144131. [CrossRef]
7. Zhai, Z.; Zhang, Y.; Cui, Y.; Zhang, Y.; Zeng, Q. Investigations on the ablation behavior of C/SiC under femtosecond laser. *Optik* **2020**, *224*, 165719. [CrossRef]
8. Zhai, Z.; Wang, W.; Zhao, J.; Mei, X.; Wang, K.; Wang, F.; Yang, H. Influence of surface morphology on processing of C/SiC composites via femtosecond laser. *Compos. Part A Appl. Sci. Manuf.* **2017**, *102*, 117–125. [CrossRef]
9. Nasiri, N.A.; Patra, N.; Ni, N.; Jayaseelan, D.D.; Lee, W.E. Oxidation behaviour of SiC/SiC ceramic matrix composites in air. *J. Eur. Ceram. Soc.* **2016**, *36*, 3293–3302. [CrossRef]
10. Yan, Z.; Lin, Q.; Li, G.; Zhang, Y.; Wang, W.; Mei, X. Femtosecond laser polishing of SiC/SiC composites: Effect of incident angle on surface topography and oxidation. *J. Compos. Mater.* **2020**, *55*, 1437–1445. [CrossRef]
11. Zhang, J.; Yuan, S.; Wei, J.; Li, J.; Zhang, Z.; Zhang, W.; Zhou, N. Spatio-temporal multi-scale observation of the evolution mechanism during millisecond laser ablation of SiCf/SiC. *Ceram. Int.* **2022**, *48*, 23885–23896. [CrossRef]
12. Wei, J.; Yuan, S.; Zhang, J.; Zhou, N.; Zhang, W.; Li, J.; An, W.; Gao, M.; Fu, Y. Removal mechanism of SiC/SiC composites by underwater femtosecond laser ablation. *J. Eur. Ceram. Soc.* **2022**, *42*, 5380–5390. [CrossRef]
13. Du, J.; Yu, G.; Jia, Y.; Ni, Z.; Gao, X.; Song, Y.; Wang, F. Ultra-high temperature ablation behaviour of 2.5 D SiC/SiC under an oxy-acetylene torch. *Corros. Sci.* **2022**, *201*, 110263. [CrossRef]
14. Liu, C.; Zhang, X.; Wang, G.; Wang, Z.; Gao, L. New ablation evolution behaviors in micro-hole drilling of 2.5D Cf/SiC composites with millisecond laser. *Ceram. Int.* **2021**, *47*, 29670–29680. [CrossRef]
15. Jiao, H.; Chen, B.; Li, S. Removal mechanism of 2.5-dimensional carbon fiber reinforced ceramic matrix composites processed by nanosecond laser. *Int. J. Adv. Manuf. Technol.* **2021**, *112*, 3017–3028. [CrossRef]

16. Gavalda Diaz, O.; Garcia Luna, G.; Liao, Z.; Axinte, D. The new challenges of machining Ceramic Matrix Composites (CMCs): Review of surface integrity. *Int. J. Mach. Tools Manuf.* **2019**, *139*, 24–36. [CrossRef]
17. Takahashi, K.; Tsukamoto, M.; Masuno, S.; Sato, Y. Heat conduction analysis of laser CFRP processing with IR and UV laser light. *Compos. Part A Appl. Sci. Manuf.* **2016**, *84*, 114–122. [CrossRef]
18. Herzog, D.; Jaeschke, P.; Meier, O.; Haferkamp, H. Investigations on the thermal effect caused by laser cutting with respect to static strength of CFRP. *Int. J. Mach. Tools Manuf.* **2008**, *48*, 1464–1473. [CrossRef]
19. Chen, X.; Li, Y.; Zhang, H.; Zhang, W. Changes in surface characteristics of stainless steel polished by nanosecond laser based on beam coupling. *Proc. SPIE* **2021**, *11907*, 1190736.
20. Zhou, Y.; Zhao, Z.; Zhang, W.; Xiao, H.; Xu, X. Experiment Study of Rapid Laser Polishing of Freeform Steel Surface by Dual-Beam. *Coatings* **2019**, *9*, 324. [CrossRef]
21. Jiang, W.; Fu, G.; Zhang, J.; Ji, S.; Shi, S.; Liu, F. Prediction of geometrical shape of coaxial wire feeding cladding in three-beam. *Infrared Laser Eng.* **2019**, *49*, 300–308.
22. Liu, F.; Ji, S.; Fu, G.; Shi, S. The Influence of Process Parameters on Geometry Characteristics by Three Beams Laser Cladding. *J. Mech. Eng.* **2020**, *56*, 227–237.
23. Shim, H.-B.; Seo, M.-K.; Park, S.-J. Thermal conductivity and mechanical properties of various cross-section types carbon fiber-reinforced composites. *J. Mater. Sci.* **2002**, *37*, 1881–1885. [CrossRef]
24. Tian, T.; Cole, K.D. Anisotropic thermal conductivity measurement of carbon-fiber/epoxy composite materials. *Int. J. Heat Mass Transf.* **2012**, *55*, 6530–6537. [CrossRef]
25. Wang, H.; Chen, X.; Zhang, W. Effects of laser beam lead angle on picosecond laser processing of silicon nitride ceramics. *J. Laser Appl.* **2019**, *31*, 042011. [CrossRef]
26. Zhang, R.; Huang, C.; Wang, J.; Zhu, H.; Liu, H. Fabrication of high-aspect-ratio grooves with high surface quality by using femtosecond laser. *Ind. Lubr. Tribol.* **2021**, *73*, 718–726. [CrossRef]
27. Zhang, X.; Chen, X.; Chen, T.; Ma, G.; Zhang, W.; Huang, L. Influence of Pulse Energy and Defocus Amount on the Mechanism and Surface Characteristics of Femtosecond Laser Polishing of SiC Ceramics. *Micromachines* **2022**, *13*, 1118. [CrossRef]
28. Chen, T.; Chen, X.; Liu, G.; Zhang, X.; Zhang, W. Research on the ablation mechanism and ablation threshold of CMC-SiC$_f$/SiC by using dual-beam coupling nanosecond laser. *Ceram. Int.* **2022**, *48*, 24822–24839.
29. Tang, Q.; Wu, C.; Wu, T. Defocusing effect and energy absorption of plasma in picosecond laser drilling. *Opt. Commun.* **2021**, *478*, 126410. [CrossRef]

 *materials*

Article

Investigation on the Coaxial-Annulus-Argon-Assisted Water-Jet-Guided Laser Machining of Hard-to-Process Materials

Yuan Li [1,†], Shuiwang Wang [1,†], Ye Ding [1,2,*], Bai Cheng [1], Wanda Xie [1] and Lijun Yang [1,2,*]

[1] School of Mechatronics Engineering, Harbin Institute of Technology, Harbin 150001, China; yuanli@hit.edu.cn (Y.L.); wsw1999@stu.hit.edu.cn (S.W.); chengbaiidx@163.com (B.C.); 13840372779@163.com (W.X.)

[2] Key Laboratory of Micro-Systems and Micro-Structures Manufacturing, Ministry of Education, Harbin Institute of Technology, Harbin 150001, China

[*] Correspondence: dy1992hit@hit.edu.cn (Y.D.); dy1992hit@163.com (L.Y.)

[†] These authors contributed equally to this work.

Abstract: In this study, the novel coaxial-annulus-argon-assisted (CAAA) atmosphere is proposed to enhance the machining capacity of the water-jet-guided laser (WJGL) when dealing with hard-to-process materials, including ceramic matrix composites (CMCs) and chemical-vapor-deposition (CVD) diamond. A theoretical model was developed to describe the two-phase flow of argon and the water jet. Simulations and experiments were conducted to analyze the influence of argon pressure on the working length of the WJGL beam, drainage circle size, and extreme scribing depth on ceramic matrix composite (CMC) substrates. A comparative experiment involving coaxial annulus and helical atmospheres revealed that the coaxial annulus atmosphere disrupts the water jet proactively, while effectively maintaining the core velocity within the confined working length and enhancing the processing capability of the WJGL beam. Single-point percussion drilling experiments were performed on a CMC substrate to evaluate the impact of machining parameters on hole morphology. The maximum depth-to-width ratio of the groove and depth-to-diameter ratio of the hole reached up to 41.2 and 40.7, respectively. The thorough holes produced by the CAAAWJGL demonstrate superior roundness and minimal thermal damage, such as fiber drawing and delamination. The average tensile strength and fatigue life of the CMCs specimens obtained through CAAAWJGL machining reached 212.6 MPa and 89,463.8 s, exhibiting higher machining efficiency and better mechanical properties compared to femtosecond (194.2 MPa; 72,680.2 s) and picosecond laser (198.6 MPa; 80,451.4 s) machining. Moreover, groove arrays with a depth-to-width ratio of 11.5, good perpendicularity, and minimal defects on a CVD diamond were fabricated to highlight the feasibility of the proposed machining technology.

Keywords: water-jet-guided laser; coaxial-annulus-argon-assisted atmosphere; machining capacity; hard-to-process materials

Citation: Li, Y.; Wang, S.; Ding, Y.; Cheng, B.; Xie, W.; Yang, L. Investigation on the Coaxial-Annulus-Argon-Assisted Water-Jet-Guided Laser Machining of Hard-to-Process Materials. *Materials* **2023**, *16*, 5569. https://doi.org/10.3390/ma16165569

Academic Editor: Norbert Robert Radek

Received: 28 July 2023
Revised: 3 August 2023
Accepted: 8 August 2023
Published: 10 August 2023

1. Introduction

The aviation and aerospace industries are experiencing rapid development, which has posed significant challenges for the performance of critical components. In response, hard-to-process materials, such as CMCs and diamond, have emerged as viable alternatives to conventional structural materials. However, owing to the distinctive mechanical, thermal, and chemical properties, these materials can hardly be processed by conventional mechanical drilling, grinding, electrical discharge machining, and laser machining with both high efficiency and quality [1–3]. Ultrafast laser machining is a proper choice, but its machining capacity is limited when dealing with a structure characterized by a large depth-to-depth/diameter ratio. Hence, it is essential to discover a general and reliable machine tool.

Water-jet-guided laser (WJGL) processing was first introduced by Richerzhagen through the approach of using a compact water jet as a multimode fiber to take advantage of both water jetting and laser machining [4]. Since its invention, this technology has found numerous applications in diverse fields, such as aviation, aerospace, semiconductor, and medical industries. For instance, Sun et al. used a WJGL to cut carbon-fiber-reinforced plastics substrate [5]. Through the multi-pass scanning, a sawtooth shape on the side wall can be observed which results from the perturbance of jet flow induced by the kerf shape. Although the formation of a heat-affected zone is suppressed, the cutting efficiency is relatively low compared with laser machining. Drilling experiments on SiC-reinforced aluminum metal matrix composites and thermal-barrier-coated alloy were put forward by Marimuthu et al. [6,7]. The holes are characterized by a good entrance circularity and low taper. By comparing the numerical and experimental results, it is indicated that when the hole depth increases, the water layer deposited at the hole bottom and substrate surface can induce the local breakup of the WJGL beam, resulting in the discontinuous interaction between the WJGL beam and the substrate, as well as a reduced material removal rate. An in-depth investigation of the surface-structure formation during the WJGL machining of a 718-nickel-based superalloy was carried out by Liao et al. [8]. Their discoveries indicate that the deposited water and confined plasma plume can trigger the appearance of recast layer, which can hardly be removed and further influences the mechanical property of the substrate.

Although WJGL has proved its feasibility in handling some materials, its machining capacity is still troubled by a typical technical bottleneck. Specifically, the water jet stability is easily interrupted by the ambient and deposited water layer, leading to the breakage of the water jet and leakage of the laser energy, which has a serious detrimental impact on the ablation process. Our previous research attempted to solve this problem by using a coaxial helical gas atmosphere [9]. The argon yields a longer working length for the WJGL beam and has a better scribing quality compared to other gases, but the machining efficiency is still limited.

In this paper, a novel coaxial annulus argon atmosphere is first introduced with the purpose of protecting the velocity core of the WJGL beam and promoting its machining capacity. The theoretical model describing the coaxial annulus argon atmosphere and water jet two-phase flow is established. The influence and related mechanism of argon pressure on the working length of the WJGL beam, drainage circle size, and extreme scribing depth on the CMC substrate are revealed, while helical atmosphere is selected as the comparison object. In this study, single-point percussion drilling experiments were carried out on a CMC substrate to uncover the influence of machining parameters on structure morphology and the extreme depth-to-diameter ratio of holes. In addition, through layer-by-layer concentric-circle drilling experiments and tensile tests on CMC specimens, the mechanical property of holes processed by the CAAAWJGL and ultrafast lasers was quantitatively evaluated. On these bases, the CAAAWJGL was applied to scribe micro grooves on CVD diamond, with a large depth-to-width ratio, good consistency, and limited defects.

2. Numerical Setup

The theoretical basis for characterizing the flow field of the CAAAWJGL is provided in the Supplementary Materials [10–12]. A three-dimensional asymmetrical model is established to demonstrate the argon–water-jet two-phase flow, as illustrated in Figure 1a. The argon phase enters the outer chamber through a 1.5 mm diameter inlet and is subsequently directed into the inner chamber through eight cylindrical channels, with a 45° angle between adjacent channels. Meanwhile, the water phase is introduced into the computational domain through a nozzle with an 80 μm diameter. Finally, the argon phase is ejected from the inner chamber, coaxially with the water jet, resulting in a two-phase argon–water-jet flow. Both the water and argon inlets are defined as pressure inlets, and the bottom of the computational domain is defined as the pressure outlet, while all other boundaries are considered to be walls. The external pressure is set at atmospheric pressure. Additionally, unstructured meshes are automatically generated using the tetra/mixed method. To ensure simulation accuracy and computational efficiency, the distance between the bottom of the inside chamber and calculation domain is set as 60 mm. The meshes surrounding the water inlet are refined due to the diameter of the water jet being within three orders of magnitude shorter than the height of the calculation domain, as shown in Figure 1b. It is important to note that the distance between the water inlet and the bottom of the inside chamber is set to 8 mm, which is consistent with the dimension in the coupling device.

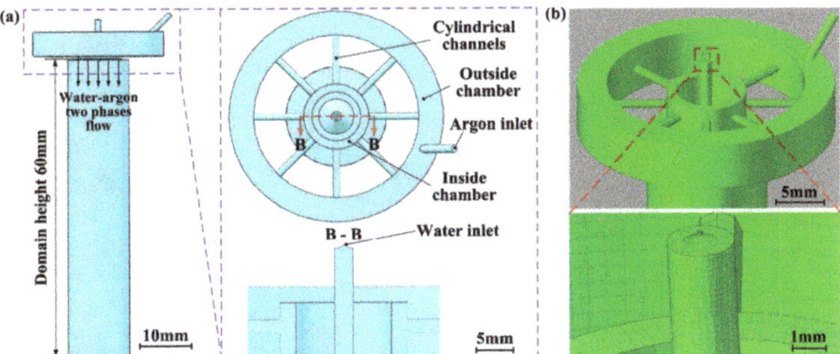

Figure 1. Calculation domain of water–argon two-phase flow: (**a**) three-dimensional structure and (**b**) its mesh generation.

The simulation process is reasonably simplified based on the following assumptions:

(1) The dynamic viscosities and densities of water and argon phases are considered to be constants.
(2) The absorption of laser energy during its transmission in the water jet is disregarded, and the internal temperature of the calculation domain is held constant at 300 K.
(3) The non-slip boundary is applied between the fluids and the wall. Additionally, the velocity gradient of fluids is assumed to be perpendicular to the wall.
(4) In the initial stage, the calculation domain is filled with ambient air, meaning that the initial volume fractions of water and argon are both zero.

3. Experimental Setup

The laser source utilized for the experiments is a Nd:YAG nanosecond laser (Pulse 532-50-LP, Inngu Laser Co., Ltd., Soochow, China), and the main output parameters are presented in Table 1. The schematic of the experimental setup for CAAAWJGL machining is demonstrated in Figure 2. After beam expansion, the laser beam is focused onto the diamond nozzle through an objective (Linos Focus-Ronar, f = 63 mm). The diameter of the laser beam is 15 mm after beam expansion. The theoretical focal spot diameter can be calculated based on Equation (1):

$$D_C = \frac{4 \cdot \lambda_L \cdot f_O}{\pi \cdot d_E} \cdot M^2 \tag{1}$$

where λ_L is the emitted laser wavelength; f_O is the focal length of the objective; d_E is laser beam diameter after expansion; M^2 is the laser-beam quality factor; and D_C is calculated as 34.13 µm, which is far smaller than the nozzle diameter (80 µm). Therefore, a well-distributed CAAAWJGL can be realized.

Table 1. Main output parameters of applied laser source.

Parameter	Wavelength	Pulse Duration	Repetition Rate	Average Power	Pulse Energy	M^2
Value	532 nm	100–120 ns	50–100 kHz	50 W (max)	1 mJ (max)	12

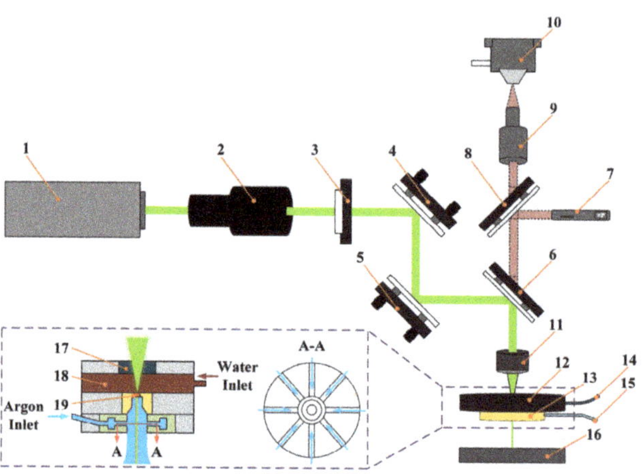

1. Laser source 2. Beam expander 3. Diaphragm 4. Reflector 1 5. Reflector 2 6.Pellicle beam splitter 1 7. Visible light source 8. Pellicle beam splitter 2 9.Attenuator 10. CCD camera 2 11. Objective 12.Coupling device 13. Argon chamber 14.Water inlet 15. Argon inlet 16. CMCs substrate 17. Optical window 18. Water chamber 19.Diamond nozzle

Figure 2. Schematic of the CAAAWJGL machining system.

The working liquid utilized in this study is deionized water (Watson). The argon gas exhibits high purity levels of 99%, with a dynamic viscosity of 22.624 µPa·s. The CMC substrates and specimens for tensile testing were obtained from the Manufacturing Technology Institute of Aviation Industry Corporation of China (Beijing, China). The CMC substrate's thickness and porosity are 3.3 mm and 6.2%, respectively. The microscopic cross-sectional morphology of the substrate is depicted in Figure 3.

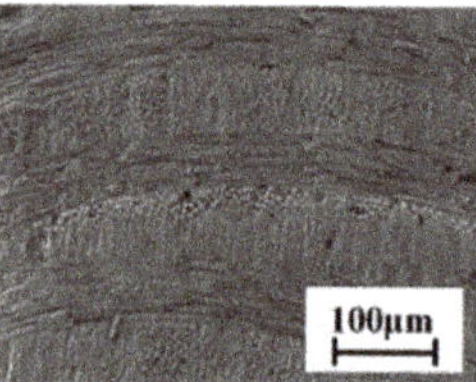

Figure 3. The cross-sectional morphology of the pristine CMC substrate.

4. Results and Discussions

4.1. The Influence of Coaxial Annulus Argon Atmosphere on WJGL Beam

Figure 4 presents the characteristics of the streamline and velocity of the argon phase within the argon chambers at inlet pressures of 20 MPa and 0.1 MPa for the water and argon phases, respectively. The streamline of water–argon two-phase flow on the central top section of reference plane α is shown in Figure 5a. Notably, despite the unilateral and oblique flow of the argon phase into the outside chamber, it exhibits a quasi-uniform flow into the inside chamber through eight cylindrical channels, as displayed in Figure 4b. Upon impinging the inner wall of the inside chamber, the velocity vector of the argon phase changes direction to become vertical, leading to a significant momentum loss, as depicted in Figure 4c. Eventually, the argon phase is ejected from the chamber in an annulus-shaped distribution that encircles the water jet, generating a shield, as shown in Figure 4d. Based on Figures 4 and 5, the maximum value of M (relative description is provided in the Supplementary Materials) is calculated as follows:

$$M_{\max} = \frac{1.784 \times 50.74^2}{997 \times 190.3^2} = 0.00013 \tag{2}$$

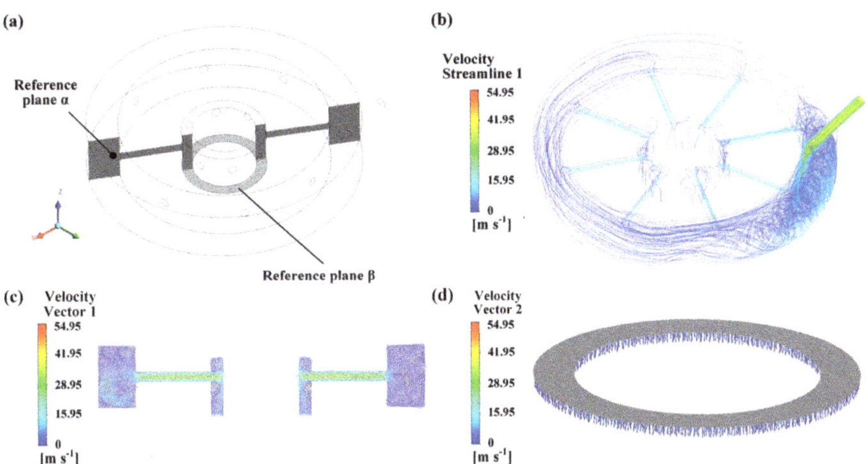

Figure 4. (a) Demonstration of reference planes α and β; (b) the overall streamline of argon, velocity vector distribution of argon on reference planes (c) α and (d) β.

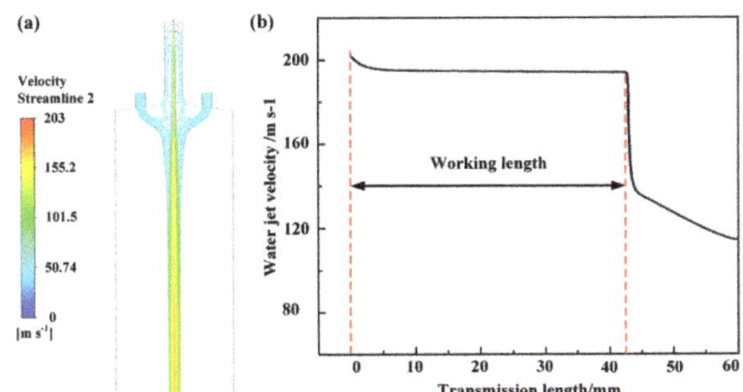

Figure 5. (a) Streamline of water–argon two-phase flow in the central top section; (b) water jet velocity along axis.

The velocity of the water jet along its central axis is displayed in Figure 5b, above, exhibiting a gradual decrease to approximately 190.3 m/s. This velocity can be maintained within the transmission length of 42.5 mm, thereby constraining the majority of the water phase within a circular region. The surrounding argon atmosphere is effectively barred from intruding into this region, enabling the laser beam to propagate along the water jet in a state of total internal reflection without any leakage. Based on the aforementioned explanations and the $M_{\max}$ value, this section constitutes the optimal zone for WJGL machining and is, hence, referred to as the working length. However, when the transmission length exceeds 42.5 mm, the water jet velocity experiences a sudden decline within 5 mm, followed by a gradual reduction. This phenomenon indicates the initiation of water jet divergence, resulting in the gradual disappearance of the jet core beyond the working length. This portion of the jet is unsuitable for WJGL machining due to its severe divergence and unstable velocity.

Figure 6 presents the working lengths of water jet under various argon pressures, as obtained from both numerical simulations and verification experiments. For the latter, a laser power of 2.2 W is used to facilitate the visualization of the working length of the water jet. Digital-camera images (captured by D90, Nikon Corporation, Tokyo, Japan) of typical morphologies are shown in Figure 7. The end of the working length is defined as the point at which the laser beam started to leak from the water jet. It should be noted that the length of the empty region on the ruler is measured as 5 mm. The working length, defined as the sum of the external stable length and the distance between the nozzle exit and bottom of the coupling device (8 mm), was used for all experiments. Each experiment was repeated five times to ensure the accuracy and reliability of the results.

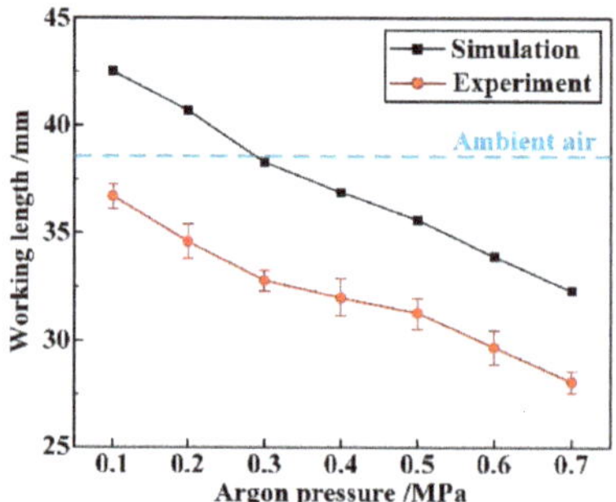

Figure 6. Comparison of working length of water jet obtained via simulations and experiments.

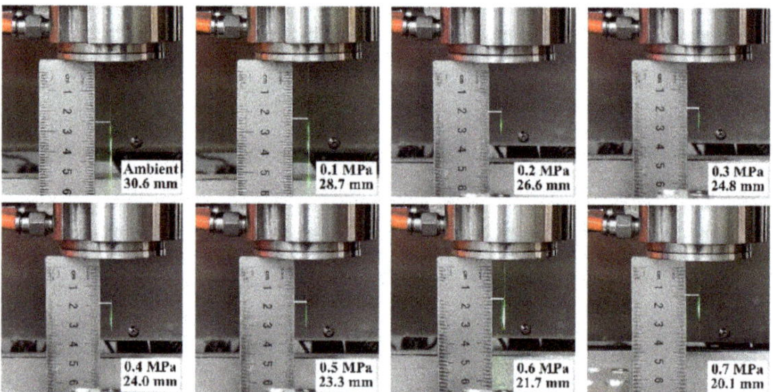

Figure 7. Typical CAAAWJGL beam morphologies and external working-length measurement results.

Referring to Figures 6 and 7, above, there exists a non-negligible difference between the numerical and experimental working lengths. This discrepancy can be attributed to the following three factors: Firstly, due to the limitations of the processing technology, the utilized diamond nozzle may exhibit defects such as notches, cracks, or burrs on its entrance and inner surface. These imperfections can trigger disturbances in the water jet as it is ejected from the nozzle. Secondly, the water in the coupling chamber is introduced by the piston pump through the high-pressure pipeline, and this process inevitably brings in impulses, causing certain turbulence. Moreover, impurities and air bubbles may be entrained into the water jet during its transmission and ejection. All of these factors contribute to the reduction in the working length of the water jet obtained from the experiments.

Furthermore, it is evident that the coaxial annulus argon atmosphere has a detrimental effect on the working length of the WJGL beam. Furthermore, this effect becomes increasingly significant with a higher argon pressure. This finding contrasts with our previous research, which utilized a helical argon atmosphere [9]. Nevertheless, it is noteworthy that the annulus argon atmosphere can ensure that the water jet maintains its core velocity, as compared to the helical atmosphere. To quantitatively characterize this phenomenon, the concentration ratio of the water jet, R_{jet}, throughout the working length is introduced in this paper, and it is defined as follows:

$$R_{jet} = \frac{D_{0.85}}{D_{jet}} \times 100\%$$
(3)

where D_{jet} is the water jet diameter at the end of working length, and $D_{0.85}$ is the circle diameter within which the water jet velocity exceeds 85% of the maximum value. The comparison of R_{jet} under coaxial annulus and helical argon atmospheres calculated by the numerical simulation is illustrated in Figure 8, with a fixed external working length and inlet water pressure of 16 mm and 20 MPa, respectively.

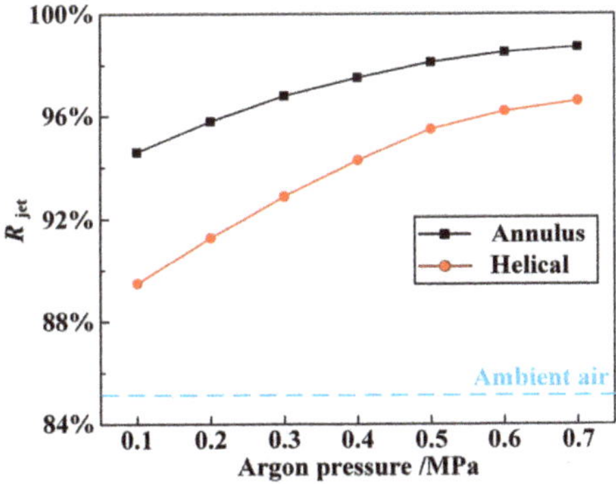

Figure 8. Comparison of R_{jet} under different argon atmospheres obtained by numerical simulations.

Verification experiments are proposed to provide insight into the aforementioned discoveries from two perspectives. Firstly, upon the incidence of the WJGL beam being put onto the substrate, an accumulated water layer manifests in the proximity of the water jet. The presence of coaxial argon atmosphere causes the rapid discharge of this water layer around the water jet, leading to the formation of a drainage circle [9], the schematic of which is demonstrated in Figure 9a. The diameters of the drainage circle under different argon atmospheres are characterized and illustrated in Figure 9b. A single-side polished silicon wafer is selected as the substrate to emphasize the morphology of the drainage circle. Additionally, single-row scribing experiments were conducted to reveal the machining capacity of the WJGL beam under different argon atmospheres. The scribing parameters are listed in Table 2. The evolution of the groove depth, as well as the morphologies of the groove at extreme depths, is demonstrated in Figure 10.

Table 2. Experimental parameters of WJGL single-row scribing.

Pulse Energy	Repetition Rate	Scribing Velocity	Scribing Time
1 mJ	50 kHz	10 mm/s	200

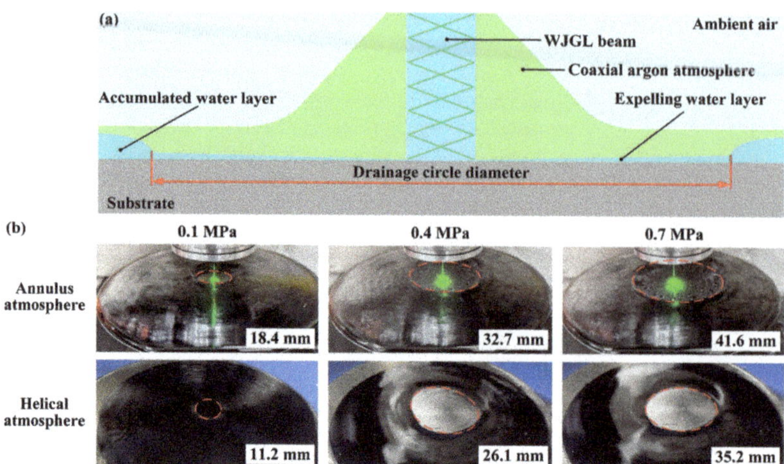

Figure 9. (**a**) Schematic of drainage circle formation. (**b**) Comparison of drainage-circle diameters under different argon atmospheres.

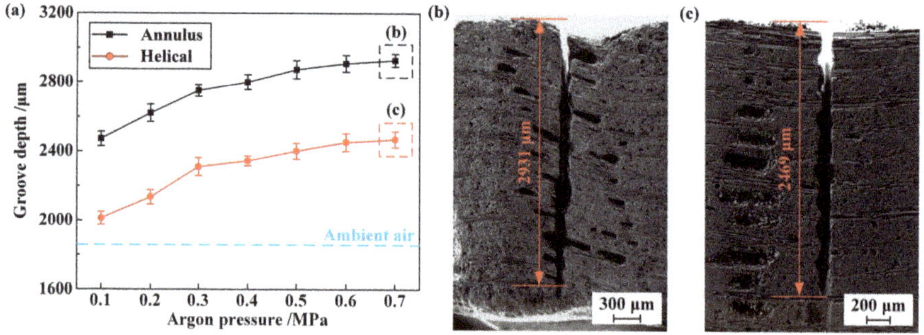

Figure 10. (**a**) Evolution of groove depth under different argon atmospheres. Cross-sectional morphology of groove at extreme depth under (**b**) annulus and (**c**) helical atmospheres.

Figure 8, above, clearly demonstrates that the assistance of annular argon atmosphere can ensure the maintenance of the core velocity of the water jet when compared with the helical atmosphere. This effect is particularly pronounced at higher argon pressures. Figures 9 and 10, above, provide further evidence of this phenomenon. In Figure 9, coaxial annular and helical argon atmospheres result in a noticeable increase in the drainage-circle diameter, with the former showing a superior performance. In Figure 10, both coaxial annular and helical argon atmospheres significantly enhance the machining capacity of the WJGL beam. To quantitatively analyze the promotion level, the promotion percentage of the groove depth, P_{gd}, is defined as follows:

$$P_{gd} = \left(\frac{H_{Ar}}{H_{air}} - 1 \right) \times 100\% \tag{4}$$

where H_{Ar} and H_{air} are the groove depth obtained under argon atmosphere and ambient air, respectively. The maximum P_{gd} under annulus and helical atmospheres is calculated as 57.84% and 32.96%, and the maximum depth-to-width ratio of the groove realized under annulus atmosphere reaches up to 41.2. Additionally, a positive correlation is observed between the promotion level and the argon pressure, but the rate of increase gradually decreased, which is similar to Figure 8.

To comprehensively analyze the numerical and experimental results, boundary layer units at the interface between different fluids near the nozzle exit were selected. The results showed that the argon phase formed a compact shield around the WLGL beam immediately after its ejection. This shield was characterized as a low swirl flow, with a maximum curl of 0.0382 in both atmospheres, based on the simulation results, indicating that the entrainment of the argon phase into the WJGL beam is negligible throughout the working length [13]. Figure 11 illustrates the velocity vectors of each phase over a very short transmission distance in the presence of different forms of argon atmosphere. Here, *V* represents the velocity; subscripts G and A represent the auxiliary argon and external air phases, respectively; and subscripts *x*, *y*, and *z* denote the direction of the velocity vector. In Figure 11a, the air phase adjacent to the WJGL beam and annulus argon shield is expelled, while the external air phase acquires velocity vectors along the Y and Z directions. On the other hand, in Figure 11b, the helical shield continuously interacts with the WJGL beam and the external air phase, leading to the external air phase acquiring velocity vectors along the X, Y, and Z directions. Since the velocities of the WJGL beam and argon near the nozzle exit are nearly identical in both atmospheres, the friction and momentum exchange between argon and the WJGL beam, as well as argon and ambient air, are much more moderate under the annulus argon atmosphere. This, in turn, results in a smaller angular deformation rate of the WJGL beam unit at the boundary layer, leading to a reduced viscous force applied to the WJGL beam unit, in accordance with the Newtonian inner friction law.

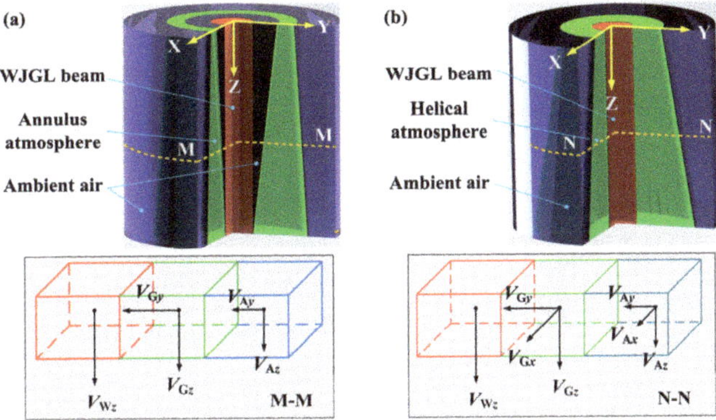

Figure 11. Schematic of velocity vector distribution at the boundary layer units between different fluids near the nozzle exit under coaxial (**a**) annulus and (**b**) helical argon atmospheres.

Therefore, the WJGL beam can maintain its initial morphology for a certain transmission length under an annular atmosphere, resulting in suppressed boundary breakage and higher jet core velocity when compared to the helical atmosphere. Consequently, there is minimal leakage of laser energy from the WJGL beam, and it impinges on the substrates at a higher velocity, which leads to a larger drainage circle [14]. These phenomena promote the machining capacity of the WJGL beam in two ways: (1) by increasing the laser energy intensity impacting the substrate and (2) by improving the water-expelling ability and reducing water deposition in the groove. The former increases the ablation volume of the CMC substrate per laser pulse, especially when the groove remains shallow. The latter provides an appropriate environment for continuous interaction between the WJGL beam and the substrate, which benefits the extension of the groove depth. This tendency is even more pronounced at higher argon pressures, as the annulus argon shield plays a more significant role in enhancing these effects.

The annulus argon atmosphere not only serves as a shield but also proactively interrupts the transmission of the WJGL beam, leading it to the transition section prematurely, as shown in Figure 12. As referenced in [15], the boundary of the WJGL beam generates a surface wave when transmitting through the gas atmosphere. The interaction between the argon and WJGL beam towards the end of the working length significantly amplifies the amplitude of the surface wave. A disturbance with a wavelength of λ_s applied to the WJGL beam boundary results in radial curvature variation, r_s, and axial curvature variation, R_s. While r_s causes the WJGL beam to diverge more rapidly, R_s restores it to a steady state. The combined effects of r_s and R_s impede the smooth transmission of the WJGL beam, leading to a notable reduction in the working length and leakage of laser energy. This phenomenon is exacerbated under higher argon pressures.

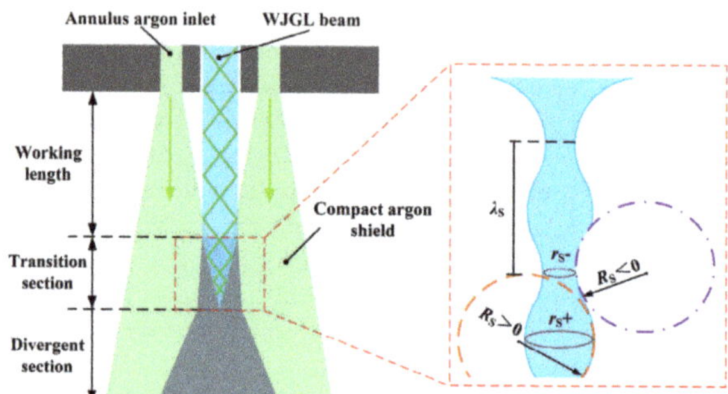

Figure 12. Schematic of WJGL beam interruption by annulus argon atmosphere.

Based on the analyses presented above, the working length of the WJGL beam is shorter under a coaxial annulus atmosphere than under a helical atmosphere. However, the WJGL beam exhibits a stronger machining capacity within this reduced length. To fully capitalize on the benefits of an annulus atmosphere, it is imperative to maintain the working length within a reasonable range.

4.2. Single-Point CAAAWJGL Percussion Drilling Experiments

According to the results in Section 4.1, a coaxial annulus argon atmosphere working at 0.7 MPa was employed to assist the WJGL beam machining of a CMC substrate. The external working length was set as 12 mm. Single-point percussion drilling experiments were conducted to investigate the influence of machining parameters on the hole morphology. The parameters are presented in Table 3. Each experiment was repeated five times. The representative morphologies of the hole entrance are depicted in Figure 13. The statistical results of the entrance and exit diameters of the holes are outlined in Figure 14.

Table 3. Experimental parameters of single-point percussion drilling.

Water Pressure/MPa	Pulse Energy/mJ	Repetition Rate/kHz	Drilling Time/s
10, 15, 20, 25, 30	0.1, 0.2, 0.4, 0.6, 0.8, 1.0	50	60

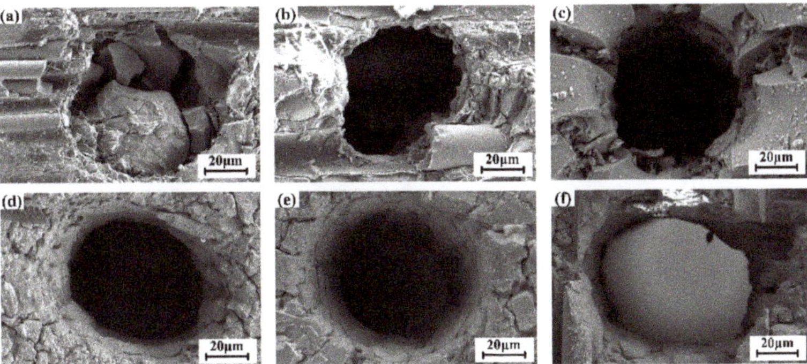

Figure 13. Entrance morphology of holes processed by percussion drilling under water pressure of 20 MPa and pulse energy of (**a**) 0.1 mJ, (**b**) 0.2 mJ, (**c**) 0.4 mJ, (**d**) 0.6 mJ, (**e**) 0.8 mJ, and (**f**) 1.0 mJ.

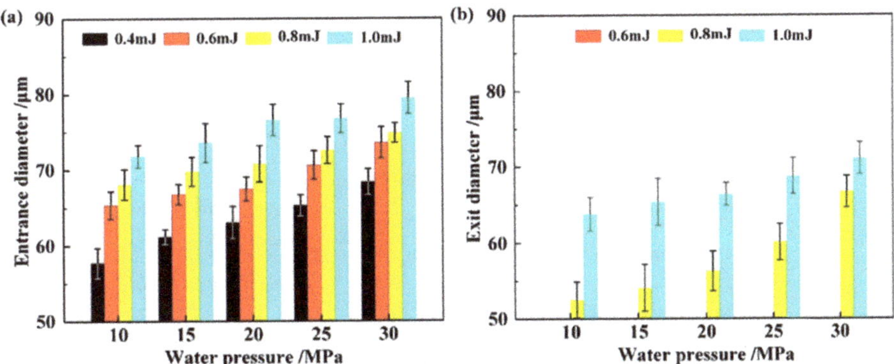

Figure 14. (**a**) Entrance and (**b**) exit diameters of holes under different water pressures and pulse energies.

Insufficient ablation occurs at the surface layer when pulse energies of 0.1 mJ and 0.2 mJ are employed, resulting in an irregular boundary of the hole entrance and unsuccessful drilling; thus, the corresponding hole diameters are excluded in Figure 14, above. To explain this phenomenon, an experimental setup was built to measure the laser energy distribution within the WJGL beam, as illustrated in Figure 15a. The WJGL beam was incident on the BK7 glass, and the laser transmitted through the BK7 glass and attenuation lens and then reached the CCD connecting to the beam quality analyzer (SP620U, Ophir-Spiricon, North Logan, UT, USA). According to Figure 15b, the utilization of low laser energy levels leads to a reduced ablation area that surpasses the substrate's ablation threshold. Consequently, the overall volume of ablation remains low, even with increasing numbers of laser pulses and water pressures. When the laser pulse energy exceeds 0.6 mJ, a thorough hole forms under all water pressures, yielding a maximum depth-to-diameter ratio of 40.7, an outcome that is difficult to achieve using other machining technologies. This finding underscores the crucial role of laser pulse energy in WJGL machining compared to other parameters. Furthermore, the maximum value of the hole entrance's diameter is smaller than that of the WJGL beam's diameter, as supported by Figure 15b. This result implies that the laser energy is constrained within the water jet during the drilling process. Consequently, the ablation is localized, reducing any extra damage on the hole's side wall.

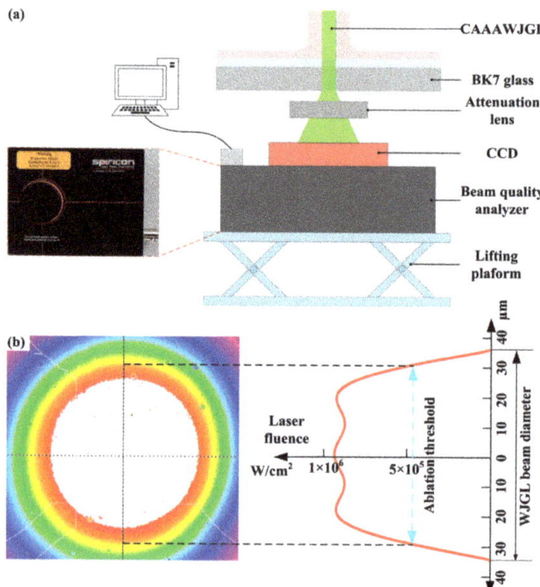

Figure 15. (**a**) Schematic of experimental setup for measuring the laser energy distribution within WJGL beam. (**b**) Measured result and schematic of ablation threshold corresponding to laser fluence.

Moreover, it is worth noting that the SiC fibers at different layers exhibit anisotropic distribution, and the diameters of warp and weft fibers in different positions also deviate from each other, as evidenced by Figure 3. When the laser beam interacts with these fibers inhomogeneously during the CAAAWJGL drilling process, the porous structure within the substrate may cause the laser beam to break up locally, resulting in the beam being reflected to an unpredictable location instead of transmitting downwards. These phenomena pose significant obstacles to achieving uniform ablation of the substrate and the formation of circular holes.

4.3. Layer-by-Layer Inside-Out CAAAWJGL Concentric-Circle Drilling and Tensile Tests

The application of CMCs in aeroengine combustors requires microhole drilling that satisfies three critical requirements: (1) a diameter in the range of hundreds of microns, (2) a good mechanical property, and (3) limited machining defects. To address these requirements, a layer-by-layer inside-out concentric-circle drilling strategy was proposed in this study [16], as shown in Figure 16a. The drilling progress at different stages is presented in Figure 16b–d, where a laser safety glass is used to enhance the clarity of the drilling process. Vertical and inclined holes were fabricated on CMC specimens, using the parameters listed in Table 4. Each experiment was repeated five times. The characteristic dimensions of the holes are listed in Tables 5 and 6. The typical entrance, exit, and cross-sectional morphologies of the holes are depicted in Figures 17 and 18.

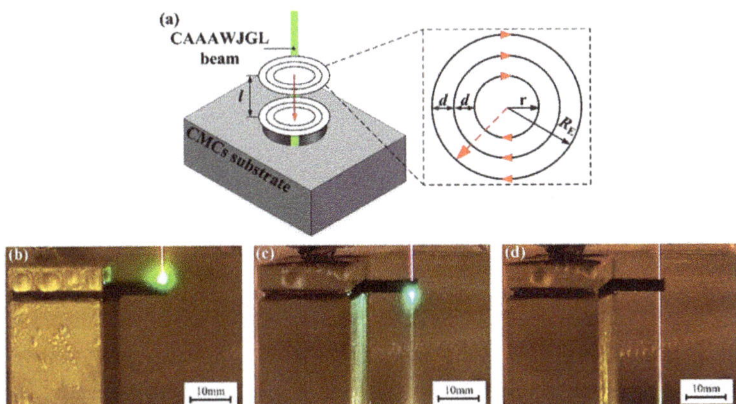

Figure 16. (**a**) Schematic of layer-by-layer inside-out concentric-circle drilling path. Digital morphology of drilling process, (**b**) drilling initialization, (**c**) hole penetration, and (**d**) end of drilling process.

Table 4. Experimental parameters of layer-by-layer inside-out concentric-circle drilling.

Parameter	$R_E/\mu m$	$d/\mu m$	$r/\mu m$	$l/\mu m$	Layer Number
Value	210	30	30	505	6
Argon pressure /MPa	Repetition rate /kHz	External working length /mm	Pulse energy /mJ	Path scanning velocity/mm·s^{-1}	Drilling time/s
0.7	50	12	1.0	10	67

Table 5. Characteristic dimensions of vertical holes.

Hole Number	1	2	3	4	5	Average
Entrance diameter/µm	493.5	487.8	491.1	490.7	488.9	490.4
Entrance circularity/µm	3.5	4.2	4.1	3.6	4.1	3.9
Exit diameter/µm	385.6	383.5	386.9	386.5	385.5	385.6
Exit circularity/µm	5.3	5.2	4.9	4.6	4.5	4.9
Taper	0.0163	0.0158	0.0158	0.0158	0.0157	0.0159
Depth-to-diameter ratio	6.69	6.77	6.72	6.73	6.75	6.73

Table 6. Characteristic dimensions of inclined holes.

Hole Number	6	7	8	9	10	Average
Entrance diameter/µm	498.5	499.3	498.1	497.9	498.7	498.5
Entrance circularity/µm	4.2	5.1	4.8	4.7	3.2	4.4
Exit diameter/µm	416.5	424.6	417.5	416.8	425.1	420.1
Exit circularity/µm	5.6	4.8	5.2	5.5	4.4	5.1
Taper	0.0124	0.0113	0.0122	0.0123	0.0112	0.0119
Depth-to-diameter ratio	6.62	6.61	6.63	6.63	6.62	6.62

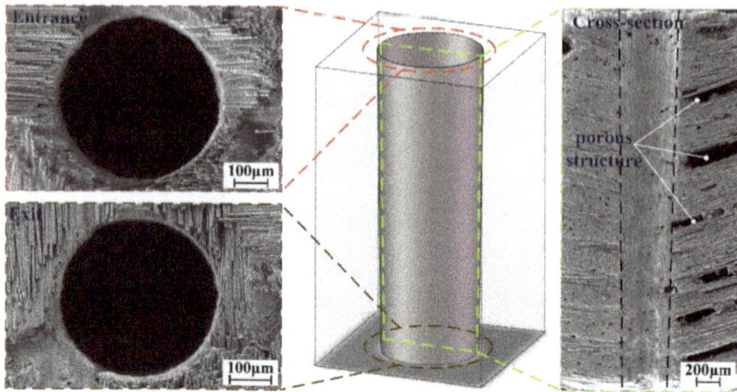

Figure 17. Typical vertical-hole morphology of CMC sample.

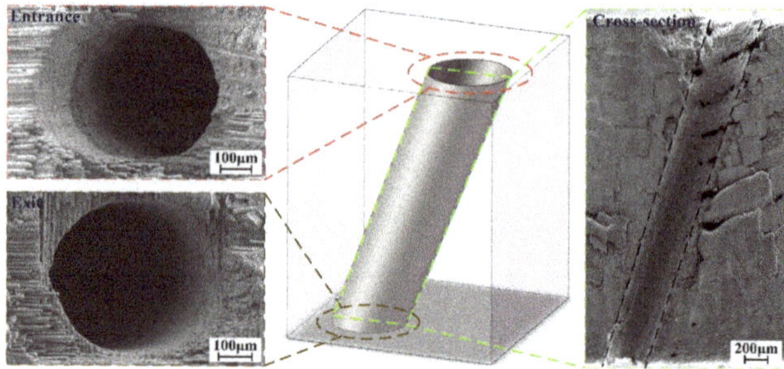

Figure 18. Typical inclined-hole morphology of CMC sample.

It is evident that both holes exhibit excellent circularity at the entrance and exit. Machining defects such as fiber drawing, microcracks, delamination, and heat-affected zones are nearly nonexistent in the cross-sectional view. Furthermore, the taper of the inclined hole is smaller than that of the vertical hole. The morphological deviation can be attributed to differences in the drainage channel. Specifically, when drilling the inclined hole, the substrate is positioned at an angle. The combined effect of gravity and the argon atmosphere facilitates the smooth expulsion of the water layer on the substrate surface and within the hole, creating an optimal environment for the effective interaction between the WJGL beam and the substrate, consequently resulting in a larger exit diameter. In contrast, when drilling the vertical hole, the drainage of the water layer relies solely on the argon atmosphere.

Tensile tests were carried out to quantitatively evaluate the mechanical property of the microhole processed by the CAAAWJGL. A thorough hole with a diameter of 0.5 mm was fabricated on the center of the specimen according to GJB 6475-2008 (a China a Chinese military standard) [17], using the parameters listed in Table 3, as illustrated in Figure 19. To highlight the advantage of CAAAWJGL machining, picosecond-laser (GX-30, Edgewave, San Diego, CA, USA) and femtosecond-laser (Pharos-15, Light Conversion, Vilnius, Lithuania) machining were chosen as the contrast objects, and the parameters are listed in Table 7. The same drilling strategy was employed. After hole drilling, the specimens were cleaned ultrasonically for 5 min and left to stand for 24 h under room temperature (300 K) and a humidity level of 65%. Then, periodic load was applied on the specimens by an electronic tensile testing machine (AGXplus, Shimadzu, Kyoto, Japan), as shown in Figure 20.

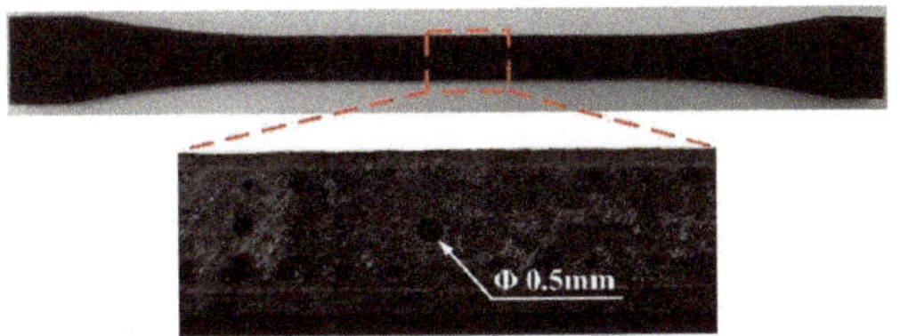

Figure 19. Morphology of CMC specimen for tensile test processed by CAAAWJGL.

Table 7. Experimental parameters of femtosecond and picosecond laser drilling.

Parameter / Laser	R_E/ Mm	d/ μm	r/ μm	l/ μm	Layer Number	Wavelength/nm
Femtosecond	230	30	20	165	20	1030
Picosecond	230	30	20	165	20	343

Parameter / Laser	Pulse duration/fs	Pulse energy /μJ	Repetition rate /kHz	Path scanning time	Path scanning velocity/mm·s^{-1}	Drilling time/s
Femtosecond	255	400	37.5	400	150	87
Picosecond	8500	75	300	1100	400	95

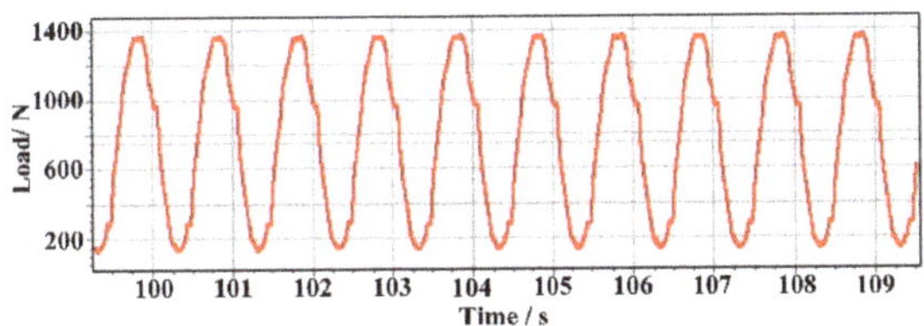

Figure 20. Periodic load applied to the CMC specimen in the tensile test.

The experimental results on the tensile strength and fatigue life of specimens processed by different machining approaches are documented in Tables 8 and 9, respectively. The characteristic morphologies of the specimens after the tensile tests are depicted in Figure 21. The results indicate that the specimens processed by CAAAWJGL exhibit a superior performance in terms of the tensile strength and fatigue life compared to those processed by ultrafast lasers. The CAAAWJGL-machined specimens demonstrate minimal instances of fiber stripping and microcracks.

Table 8. Tensile strength of CMC specimens processed by different machining approaches.

Machining Approach	Specimen 1/MPa	Specimen 2/MPa	Specimen 3/MPa	Specimen 4/MPa	Specimen 5/MPa	Average /MPa
CAAAWJGL	198	237	209	214	205	212.6
Femtosecond	212	182	177	202	198	194.2
Picosecond	183	217	206	185	202	198.6

Table 9. Fatigue life of CMC specimens processed by different machining approaches.

Machining Approach	Specimen 1	Specimen 2	Specimen 3	Specimen 4	Specimen 5	Average
CAAAWJGL/s	90,854	87,621	91,215	89,541	88,448	89,463.8
Femtosecond/s	74,510	71,362	70,651	73,021	73,857	72,680.2
Picosecond/s	79,954	81,825	77,514	78,659	80,305	80,451.4

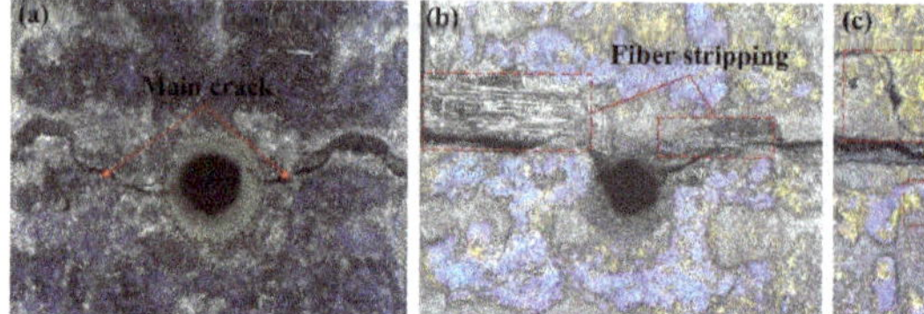

Figure 21. Characteristic specimen morphologies after tensile tests processed by (**a**) CAAAWJGL, (**b**) femtosecond laser, and (**c**) picosecond laser.

Based on the experimental outcomes and discoveries in the existing literature [18], the CAAAWJGL machining technique demonstrates its superiority through its stronger machining capacity and better mechanical property compared to dry ultrafast laser machining. The reasons for this can be elucidated as follows.

The CAAAWJGL machining process involves the simultaneous actions of scouring and cooling via the modified water jet, whereby the debris and confined plasma plume are swiftly expelled by the rapid water jets subsequent to its generation, avoiding the formation of molten deposits on the machined surface. Consequently, the boundary between warp and weft SiC fiber is distinct without any drawing, thereby preserving the mechanical properties of the machined surface. Additionally, the CAA atmosphere effectively removes the accumulated water layer at both the structure bottom and surface, creating a clean environment for the subsequent machining process. Consequently, the WJGL beam can continuously ablate the substrate, ensuring a constant machining capacity throughout the machining process.

In contrast, the dry ultrafast laser machining relies on auxiliary gas to expel the debris, thus demonstrating considerably lower efficacy compared to the high-speed water jet, and it is prone to inducing residual deposition, fiber drawing, and an oxide layer. These machining defects significantly impact the mechanical properties of the machined surface [19]. Moreover, the shielding effect triggered by laser-induced plasma further sets up obstacles for the interaction between laser and substrate, which can hardly be suppressed by the auxiliary gas and leads to the sharp shrinkage of the machining capacity.

4.4. Further Application of CAAAWJGL on Scribing CVD Diamond

In order to extend the application of CAAAWJGL machining to the aviation and aerospace industries, scribing experiments were conducted on 3.5 mm thick CVD diamond (Shanghai Academy of Spaceflight Technology, Shanghai, China). This material is currently employed in satellite heat sinks and is acknowledged as one of the most challenging materials to machine. The scribing strategy is based on our previous research [20], and the scribing parameters are listed in Table 10. The groove morphologies are depicted in Figure 22 and Table 11, demonstrating good perpendicular sidewalls without any signs of turning. Notably, the grooves exhibit a depth-to-width ratio of 11.5, with an average width and depth of 175.8 μm and 2030 μm, respectively. Moreover, typical defects, such as microcracks, edge breakage, carbonization, and large taper, are absent. These results suggest that none of the existing machining technologies is capable of achieving such high-quality and high-efficiency machining of CVD diamond.

Table 10. Experimental parameters of grooves scribing on CVD diamond.

Water Pressure/MPa	Argon Pressure /MPa	Repetition Rate /kHz	External Working Length /mm	Scribing Interval on the Same Layer/μm	Scribing Layer along Depth	Pulse Energy /mJ	Path Scanning Velocity/mm·s^{-1}
30	0.7	50	16	20	50	1.0	20

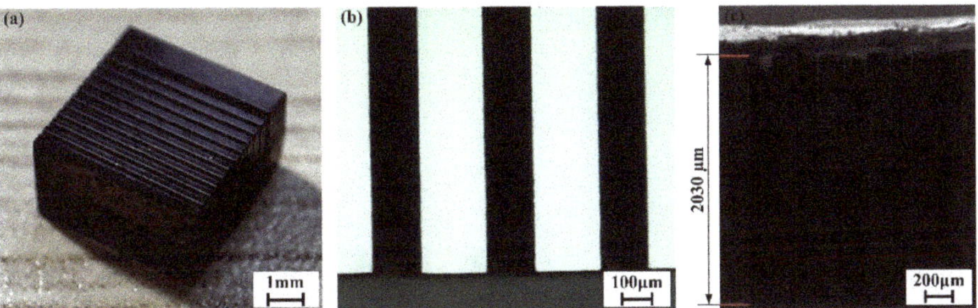

Figure 22. (**a**) The overall, (**b**) top, and (**c**) cross-sectional morphology of grooves on CVD diamond.

Table 11. Characteristic dimensions of grooves on CVD diamond.

Groove Number	1	2	3	4	5	Average
Width/μm	177.8	173.4	173.7	176.8	177.3	175.8
Depth/μm	2030.5	2042.6	2035.8	2032.2	2008.9	2030.0

5. Conclusions

In this study, a coaxial annulus argon atmosphere was proposed to assist the WJGL machining of hard-to-process materials with high quality and efficiency, including CMCs and CVD diamond. The discoveries can be summarized as follows:

(1) The introduction of coaxial annulus argon atmospheres proactively interrupts the water jet, leading to a decrease in the working length. Nevertheless, compared to the utilization of a coaxial helical atmosphere, the CAAAWJGL beam exhibits superior maintenance of its core velocity within the constrained working length, therefore enhancing the processing capability. This enhancement is manifested in a single row scribing depth-to-width ratio of 41.2, representing a noteworthy improvement of 57.84% and 32.96% compared to ambient conditions and a helical atmosphere, respectively.

(2) The modification of the CAAAWJGL beam improves the ablation capacity of single-point percussion drilling on CMCs, obtaining a maximum depth-to-diameter ratio of 40.7. Furthermore, by employing an inside-out strategy, the successful fabrication of vertical and inclined holes is accomplished with exceptional circularity at both the entrance and exit, minimal taper, and negligible fiber drawing and delamination. The CAAAWJGL exhibits superior processing efficiency compared to femtosecond and picosecond lasers, meanwhile achieving a better tensile strength and fatigue life. Furthermore, the feasibility of CAAAWJGL machining is demonstrated through the scribing of CVD diamond with good perpendicularity and minimal defects on CVD diamond.

However, it should be noted that the reduced working length limits the machining flexibility when dealing with components featuring complex curved surface features. Future research will focus on exploring approaches to overcome this technical bottleneck, such as optimizing the gas chamber's structure and atmosphere composition.

Supplementary Materials: The following supporting information can be downloaded at https://www.mdpi.com/article/10.3390/ma16165569/s1, Theoretical basis for characterizing the flow field of CAAAWJGL [10–12,21,22].

Author Contributions: Conceptualization, S.W. and W.X.; Methodology, Y.D.; Validation, B.C.; Formal analysis, L.Y.; Investigation, Y.L.; Data curation, W.X.; Writing—original draft, Y.L. and S.W.; Supervision, Y.D. and L.Y. All authors have read and agreed to the published version of the manuscript.

Funding: This research was supported by the National Major Science and Technology Project of China (No. 2019-VII-0009-0149), the National Key Research and Development Program of China (No. 2021YFF0500200), and the Science Center for Gas Turbine Project (No. P2022-A-IV-002-003).

Institutional Review Board Statement: Not applicable.

Informed Consent Statement: Not applicable.

Data Availability Statement: The data presented in this study are available on request from the corresponding author.

Conflicts of Interest: The authors declare that they have no known competing financial interest or personal relationships that could have appeared to influence the work reported in this paper.

References

1. Doi, T. Next-generation, super-hard-to-process substrates and their high-efficiency machining process technologies used to create innovative devices. *Int. J. Autom. Technol.* **2018**, *12*, 145–153. [CrossRef]
2. Axinte, D.; Billingham, J.; Bilbao, G.A. New models for energy beam machining enable accurate generation of free forms. *Sci. Adv.* **2017**, *3*, e1701201. [CrossRef]
3. Lu, M.Y.; Zhang, M.; Zhang, K.H.; Meng, Q.G.; Zhang, X.Q. Femtosecond UV laser ablation characteristics of polymers used as the matrix of astronautic composite material. *Materials* **2022**, *15*, 6771. [CrossRef]
4. Richerzhagen, B.; Kutsuna, M.; Okada, H.; Ikeda, T. Water-jet-guided laser processing. *Proc. SPIE* **2003**, *4830*, 91–94. [CrossRef]
5. Sun, D.; Han, F.Z.; Ying, W.S. The experimental investigation of water jet–guided laser cutting of CFRP. *Int. J. Adv. Manuf. Technol.* **2019**, *102*, 719–729. [CrossRef]
6. Marimuthu, S.; Dunleavey, J.; Liu, Y.; Smith, B.; Kiely, A.; Antar, M. Water-jet guided laser drilling of SiC reinforced aluminium metal matrix composites. *J. Compos. Mater.* **2019**, *53*, 3787–3796. [CrossRef]
7. Marimuthu, S.; Smith, B. Water-jet guided laser drilling of thermal barrier coated aerospace alloy. *Int. J. Adv. Manuf. Technol.* **2021**, *113*, 177–191. [CrossRef]
8. Liao, Z.; Xu, D.; Axinte, D.; Diboine, J.; Wretland, A. Surface formation mechanism in waterjet guided laser cutting of a Ni-based superalloy. *CIRP Ann.* **2021**, *70*, 155–158. [CrossRef]
9. Cheng, B.; Ding, Y.; Li, Y.; Li, J.Y.; Xu, J.J.; Li, Q.; Yang, L.J. Coaxial helical gas assisted laser water jet machining of SiC/SiC ceramic matrix composites. *J. Mater. Process. Technol.* **2021**, *293*, 117067. [CrossRef]
10. Nektarios, K.; John, G.B.; Nicolas, C.M. Evaluation of Reynolds stress, k-ε and RNG k-ε turbulence models in street canyon flows using various experimental datasets. *Environ. Fluid Mech.* **2012**, *12*, 379–403. [CrossRef]
11. Yakhot, V.; Orszag, S.A.; Thangam, S.; Gatski, T.B.; Speziale, C.G. Development of turbulence models for shear flows by a double expansion technique. *Phys. Fluids A Fluid Dyn.* **1992**, *4*, 1510–1520. [CrossRef]
12. Lasheras, J.C.; Hopfinger, E.J. Liquid jet instability and atomization in a coaxial gas stream. *Annu. Rev. Fluid Mech.* **2000**, *32*, 275–308. [CrossRef]
13. Clanet, C.; Lasheras, J.C. Depth of penetration of bubbles entrained by a plunging water jet. *Phys. Fluids* **1997**, *9*, 1864–1866. [CrossRef]
14. Chigier, N.A.; Beer, J.M. Velocity and static pressure distributions in swirling air jets issuing form annular and divergent nozzles. *J. Basic Eng.* **1964**, *86*, 788–798. [CrossRef]
15. Yang, M.G.; Gong, C.; Wang, Y.L.; Lu, J.G. Experimental study of ultra-high pressure capillary water jet based on image intensity. *J. Propuls. Technol.* **2014**, *35*, 87–92. [CrossRef]
16. Yang, L.J.; Ding, Y.; Wang, M.L.; Cao, T.T.; Wang, Y. Numerical and experimental investigations on 342 nm femtosecond laser ablation of K24 superalloy. *J. Mater. Process. Technol.* **2017**, *249*, 14–24. [CrossRef]
17. Available online: https://www.antpedia.com/standard/5788433.html (accessed on 7 August 2023).
18. Oriol, G.D.; Gonzalo, G.L.; Liao, Z.R.; Dragos, A. The new challenges of machining ceramic matrix composites (CMCs): Review of surface integrity. *Int. J. Mach. Tools Manuf.* **2019**, *139*, 24–36. [CrossRef]
19. Zhai, Z.Y.; Wei, C.; Zhang, Y.C.; Cui, Y.H.; Zeng, Q.R. Investigations on the oxidation phenomenon of SiC/SiC fabricated by high repetition frequency femtosecond laser. *Appl. Surf. Sci.* **2020**, *502*, 144131. [CrossRef]
20. Li, Y.; Ding, Y.; Cheng, B.; Cao, J.J.; Yang, L.J. Numerical and experimental research on the laser-water jet scribing of silicon. *Appl. Sci.* **2022**, *12*, 4057. [CrossRef]
21. Villermaux, E.; Rehab, H.; Hopfinger, E.J. Breakup regimes and self-sustained pulsations in coaxial jets. *Meccanica* **1994**, *29*, 393–401. [CrossRef]
22. Rehab, H.; Villermaux, E.; Hopfinger, E.J. Flow regimes of large-velocity-ratio coaxial jets. *J. Fluid Mech.* **1997**, *345*, 357–381. [CrossRef]

Article

Research on Electrostatic Field-Induced Discharge Energy in Conventional Micro EDM

Yaou Zhang [1,2,*], Qiang Gao [1,2], Xiangjun Yang [1,2], Qian Zheng [1,2] and Wansheng Zhao [1,2]

[1] State Key Laboratory of Mechanical System and Vibration, Shanghai 200240, China; 121020920638@sjtu.edu.cn (X.Y.); zhengqian@sjtu.edu.cn (Q.Z.)

[2] School of Mechanical Engineering, Shanghai Jiao Tong University, Shanghai 200240, China

* Correspondence: yaou_zhang@sjtu.edu.cn; Tel./Fax: +86-21-3420-6951

Abstract: The electrostatic field-induced electrolyte jet (E-Jet) electric discharge machining (EDM) is a newly developed micro machining method. However, the strong coupling of the electrolyte jet liquid electrode and the electrostatic induced energy prohibited it from utilization in conventional EDM process. In this study, the method with two discharge devices connecting in serials is proposed to decouple pulse energy from the E-Jet EDM process. By automatic breakdown between the E-Jet tip and the auxiliary electrode in the first device, the pulsed discharge between the solid electrode and the solid workpiece in the second device can be generated. With this method, the induced charges on the E-Jet tip can indirectly regulate the discharge between the solid electrodes, giving a new pulse discharge energy generation method for traditional micro EDM. The pulsed variation of current and voltage generated during the discharge process in conventional EDM process verified the feasibility of this decoupling approach. The influence of the distance between the jet tip and the electrode, as well as the gap between the solid electrode and the work-piece, on the pulsed energy, demonstrates that the gap servo control method is applicable. Experiments with single points and grooves indicate the machining ability of this new energy generation method.

Keywords: field-induced discharge; energy generation method; electrolyte jet electrode; discharge in gas; micro EDM

Citation: Zhang, Y.; Gao, Q.; Yang, X.; Zheng, Q.; Zhao, W. Research on Electrostatic Field-Induced Discharge Energy in Conventional Micro EDM. *Materials* **2023**, *16*, 3963. https://doi.org/10.3390/ma16113963

Academic Editor: Andrea Gatto

Received: 18 April 2023
Revised: 18 May 2023
Accepted: 23 May 2023
Published: 25 May 2023

1. Introduction

Micro EDM (electrical discharge machining) is a specialized machining method that processes conductive materials by using the thermal energy produced by pulse discharges between two electrodes, which melts and vaporizes the material from the workpiece as well as the tool electrode. This method yields numerous benefits. For example, EDM does not make direct contact between the electrode and the workpiece eliminating mechanical stresses, chatter, and vibration problems during machining [1], and the hardness of the tool electrode material may be lower than that of the workpiece material. It can process materials that are difficult to deal with by traditional methods [2], and is particularly suited for small and complex features in aviation [3], diesel engines [4], nuclear power plants [5], and medical [6] and mold [7] manufacturing industries.

The discharge energy is significant because it is intimately related to material removal rate, machining accuracy, and surface integrity. In micro EDM, the small pulse energy is one of the prominent research topics. The major discharge energy generation methods are the relaxation pulse generator [8] and the transistor-pulse generator [9]. Singh et al. [8] verified that the relaxation pulse generator can generate low discharge energy by charging and discharging the capacitor adopted in the main power loop. Ichikawa and Natsu [10] further confirmed that the existing stray capacitance in the machine tool limited its access to the minimal discharge energy with a short duration and high frequency, even if removing the capacitors in the main power loop. That is, the ineliminable stray capacitance plays a key role in reducing discharge energy. Ashwin and Muthuramalingam [9] claimed that a transistor-type pulse generator can generate high-frequency discharge energy. Chu et al. [11] pointed out that the device response time spent in intricate detection, power amplification, and complex control circuits in transistor-type pulse generator prevents the pulse energy from further reducing. In principle, the existing pulse power generation methods all have unavoidable problems in reducing the discharge energy. As a result, new discharge energy generation theory has to be studied to reduce the discharge energy per pulse to meet the needs of micro-EDM. On the other hand, high-speed rotation of the tool electrode is commonly required in micro-hole machining to improve debris removal and assure the hole's roundness. The brush must be used to connect the stationary pulse power supply to the rotatory tool electrode [12]. The delicate brush might be abraded or destroyed due to the direct contact of pulse-type electricity to the conductive main spindle. The greater the frequency and rotation speed, the larger the brush loss. This brings hidden dangers to the use of high-frequency pulse in micro EDM. As a result, researchers began to focus on new pulse power supply generation theory and the brushless energy transfer method in micro EDM.

In terms of new discharge energy generation methods, Abbas et al. [13] and Koyano et al. [14] presented a structure with a feed capacitor linked in series with a circuit consisting of a pulse power supply, tool electrode, and workpiece. When the pulse power supply is turned on, induced charges can be induced on the feed capacitor and the tool electrode and the workpiece simultaneously. Breakdown and discharge can occur if the tool electrode and the workpiece are sufficiently near to each other at this time. Then, the pulse power supply's two ends are immediately connected, allowing the feed capacitor to serve as the pulse power supply, resulting in the reverse breakdown discharge between the workpiece and tool electrode [13]. Furthermore, Yahagi et al. [15] employed a 50,000 rpm spindle electrode as one plate of the feed capacitor, enabling the induced brushless power supply. However, according to Zou et al. [16], the polarity effect dominates the micro-EDM process. Although bipolar power supply machining improves deionization and discharge stability [13], it is still detrimental to machining efficiency and electrode wear. Furthermore, the feeding capacitor and the inter-electrode discharge gap must be precisely regulated, as must the matching of the pulse on and short-circuit sequencing control across the power supply.

The electrostatic field-induced electrolyte jet EDM was proposed by Zhang et al. [17]. This approach removes material by plasma breakdown of a high-voltage DC electrostatic field between the flexible jet electrode and the workpiece, and utilizes the induced charges on the jet tip and workpiece as discharge energy. Zhang et al. [18] also confirmed that by modifying the electrolyte concentration and electric parameters, the single pulse energy can be lowered to less than 10×10^{-6} J. However, this pulse energy is difficult to transfer to conventional EDM since it is tightly coupled with the jet electrode generation process.

This work proposes a structure and tries to decouple this energy from the jet electrode to apply it into conventional micro EDM based on previous research [17,18]. The paper is organized as follows: Section 2 constructs a decouple test platform; Section 3 explains this decouple theory and the working principle of the machining process as well as the equivalent circuit. The experimental results reveal the influence law of the E-Jet electrode discharge on the discharge of the solid electrode side. The single-point and groove machining experiments in conventional EDM with the proposed method are completed to validate its machining ability. Finally, the contents are summarized and conclusions are drawn in the final section.

2. Construction of Test Platform

An experimental platform is established to apply the E-Jet discharge energy generation method into conventional micro EDM, as shown in Figure 1. This platform is primarily made up of four components: the E-Jet EDM component, the traditional EDM component, the current and voltage sensing component, and the workpiece. This workpiece is made up of two parts: the sacrificial one and the right one to be machined, which are joined together by a conductive block. The first E-Jet EDM section is on the left side of the workpiece and consists primarily of a syringe with a flat pipe-like spray nozzle filled with the electrolyte, as well as a 3D manual motion platform (A) used to adjust the position of the nozzle relative to the sacrificial work-piece. One terminal of the high-voltage DC power supply is connected to the syringe's metallic needle. The conventional EDM section is on the right side and is nearly a mirror image of the first component. The only difference lies in that the syringe and flat-headed nozzle do not contain any electrolytes. Instead, a solid tungsten needle with a diameter of 0.3 mm is placed inside the nozzle, extending out as the solid electrode. By adjusting the 3D motion platform (B), the gap between the solid electrode and the workpiece can be regulated. On this side, the other end of the high-voltage DC power supply is connected to the solid tungsten needle. In the experiment, Spellman SL60 (Spellman High Voltage Electronics Corporation, Hauppauge, NY, USA) module is selected as the high-voltage DC power supply. The third sensing component, consists of two differential voltage probes and a current sensor. The current sensor (Tektronix current sensor TCP0030A, TEKTRONIX INC., Beaverton, OR, USA) measures the discharge current, and the voltage between the electrostatic field-induced jet and the sacrificial electrode, as well as the voltage between the solid electrode and the workpiece, is measured by two Tektronix high voltage differential probes (P6015A, TEKTRONIX INC., Beaverton, OR, USA). During the discharge process, the Tektronix oscilloscope (MDO3104, TEKTRONIX INC., Beaverton, OR, USA) is used to collect current and voltage signals online.

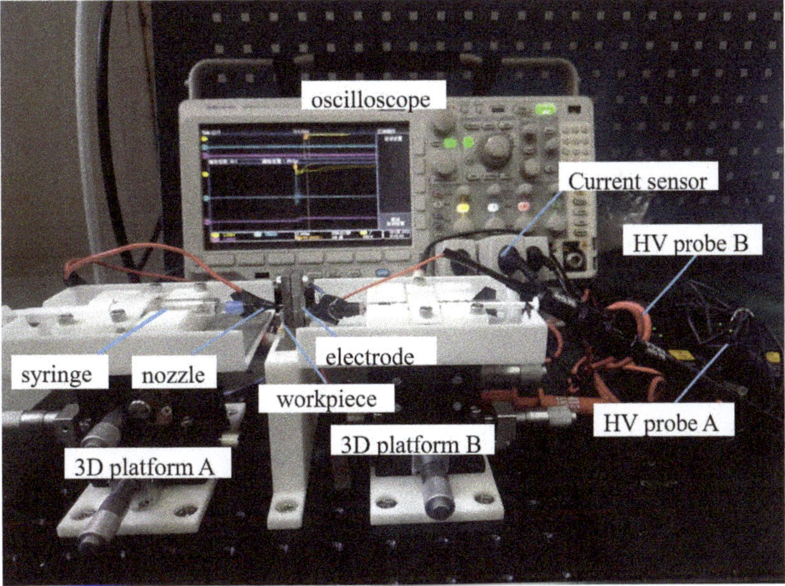

Figure 1. The experimental platform of the E-Jet discharge energy used in traditional EDM.

The schematic representation of the experimental apparatus is shown in Figure 2. The power supply is a high-voltage DC with a voltage range of 0–10 kV. It can generate a strong electric field, which drags the electrolyte jet from the flat-headed nozzle toward the sacrificial electrode without any pump [18]. With the electrolyte jet extending, the field intensity increases and results in the plasma breakdown. Due to the series connection, the discharges between the solid electrode and the workpiece and between the E-Jet and workpiece almost occur in the meantime. Materials are removed on both sides of the workpiece. Following the discharge, the surface tension can pull back the electrolyte jet towards the nozzle, leading to the discharge interruption and forming a pulse interval. As a result, periodic spontaneous discharges can occur.

On the E-Jet EDM side, the discharge energy is generated by the induced charges on the jet surface and the workpiece. The dynamic imbalance of the electric force of the induced charges and electrolyte surface tension produces the electrostatic-induced pulsed jet, which is used as the tool electrode. As a result, the generation of jet electrodes and that of discharge energy are always tightly coupled. This E-Jet EDM varies greatly from electrolyte jet machining (EJM), which uses a high-pressure pump to force electrolytes onto components to remove material by electrochemical machining, as described by Liu et al. [19]. Furthermore, it is also different from the electrochemical discharge machining (ECDM) proposed by Goud et al. [20] and Singh and Dvivedi [21], in which bubbles generated by electrochemical reactions gather around the tool electrode, and the material is etched by the discharge generated by the breakdown of the gas bubble.

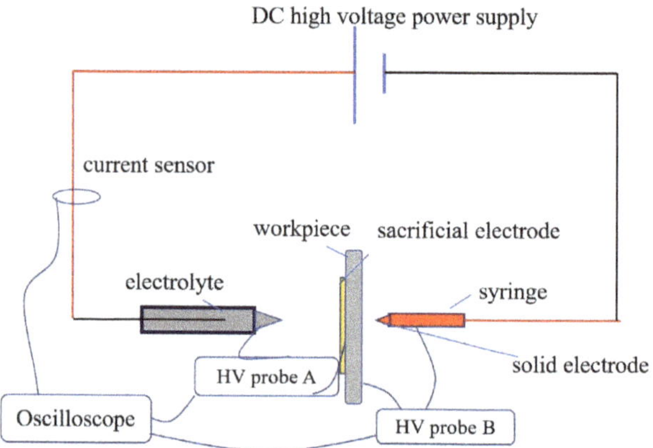

Figure 2. Schematic diagram of energy generation principle of micro EDM based on E-jet process.

In Figure 2, the pulsating discharge of the solid electrode and the workpiece in conventional EDM can be controlled by the dynamical E-Jet EDM by joining the sacrificial electrode to the workpiece. The pulsating discharge of the E-Jet EDM can be used to control the traditional EDM process. The power source is a high-voltage DC voltage power supply without any external other pulse power supply or pulse control circuit, and the spontaneous periodic E-Jet discharge process can realize the periodic discharge on traditional EDM side.

3. Experimental Results and Discussion

3.1. Equivalent Circuit Analysis of the Discharge Process

The electrolyte in the nozzle is connected to the positive terminal of the high-voltage DC power supply in Figure 2, and the solid electrode is connected to the negative terminal. The sacrificial workpiece and the connected workpiece between the nozzle and the solid electrode are suspended in potential.

At first, the jet dragged from the nozzle is induced with positive charges on its surface by the high-voltage DC electrostatic field, while the solid electrode induces with negative charges. As shown in Figure 3a, the negative charges are induced on the surface of the sacrificial workpiece opposite the jet end, while positive charges are induced on the surface of the workpiece opposite the solid electrode. The total amount of generated positive and negative charges on these two sides of workpiece should be identical, but the distributions are different in space and polarity. The micro-E-Jet is emitted from the cone tip when the electric field force of induced charges on the Taylor cone at the nozzle outlet exceeds the liquid surface tension, which is similar to the phenomenon observed by Rosell-Llompart et al. [22], causing the gap between the jet tip and the workpiece to narrow. The E-Jet discharges with the sacrificial workpiece because the electric field strength between the E-Jet tip and the workpiece is larger than the air breakdown voltage [17], as illustrated in Figure 3b. The jet and plasma appear to be connecting the nozzle to the workpiece at this moment. The power supply voltage is then delivered between the solid electrode and the workpiece, causing the charge distribution on the workpiece surface to rapidly gather. As shown in Figure 3c, the inter-electrode voltage rises sharply, causing discharge breakdown between the solid electrode and the workpiece, resulting in traditional EDM in gas. At that point, both the discharges between the E-Jet and the workpiece and between the solid electrode and the workpiece are active. If the plasma impedance is ignored, it can be treated as if two ends of the power supply are connected. The discharge reduces the induced charges on the jet surface quickly. The jet retreats to the nozzle outlet as the electric field forces on it decrease and become less than the surface tensions. As shown in Figure 3d, the discharge between the solid electrode and the workpiece begins to extinguish at this point. Soon after, the cone, solid electrode, and surfaces on both sides of the workpiece resume charging in the strong electric field. As a result, the electrostatic field-induced jet, solid electrode, and workpiece surface discharge spontaneously to form a periodic continuous discharge process. As a consequence, the usage of the E-Jet EDM energy generating technology in conventional EDM can be realized indirectly. Because the field jet's discharge energy may potentially be regulated to a very tiny scale [18], it can circumvent the discharge energy constraint in conventional micro EDM with the relaxation pulse generator [10] and the transistor-pulse generator [11]. This technology may provide an innovative method of pulsating discharge energy to micro EDM. Furthermore, this non-contact indirect electrostatic induction method provides a new solution to brushless energy transmission to high-speed spindles in micro-EDM.

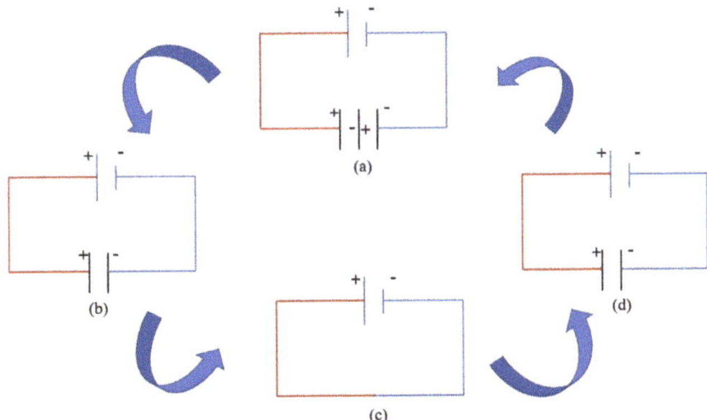

Figure 3. Decomposition schematic diagram of E-Jet energy generation method in micro EDM: (**a**) the initial electric field distribution state; (**b**) the equivalent electric field distribution after an E-Jet discharge generation; (**c**) the equivalent electric field distribution when both sides of dis-charges are active; (**d**) the equivalent electric field distribution when the E-jet retreats with only solid electrode discharge remaining.

3.2. Factors Affecting Discharge Energy

Before using the field-induced jet discharge energy, the voltages and currents on the E-Jet electrode side and solid electrode side should be considered. Moreover, to disclose the induced discharge mechanism of the solid electrode side and verify the equivalent circuit built in the last section, the sequency of these two discharges should be taken into consideration. Furthermore, to control the EDM process, the distance between the solid electrode and the workpiece, as well as the distance between the E-Jet electrode and the workpiece, should be studied, which can build a foundation for gap servo control during the EDM process.

3.2.1. Discharge Waveforms Analysis

The key to distinguishing these serials' structure discharges process is to study the waveforms of currents and voltages. To observe these discharges, a higher open circuit voltage is used to make them more visible and severe. During the discharge process, the voltage between the E-Jet nozzle and the sacrificial electrode has experienced a sharp breakdown drop to the lowest point, a stable maintaining discharge process, a slow charging rise, and a rapid exponential charging rise to the open circuit voltage, as shown in Figure 4. However, the voltage between the solid electrode and the workpiece rises from zero to 400 V and then gradually falls back to zero. The inter-electrode current rises sharply during the discharge period, then decreases and stays in a plateau for a short period before returning to zero after the discharge.

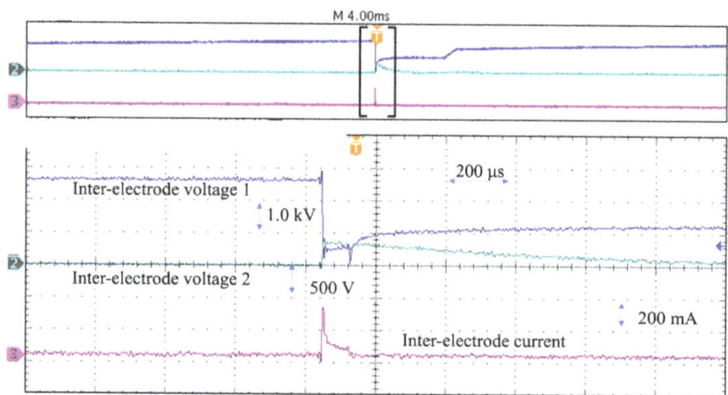

Figure 4. The voltages and currents waveforms during the discharge processes.

When compared to the experimental results of Ashwin and Muthuramalingam [9] and Kunieda et al. [23], this discharge current waveform appears to be a conventional pulse discharge waveform. Because the breakdown involves a time lag in which the initial electrons are accelerated to form the electron avalanche, Kunieda et al. [23] noticed that the discharge happens after the ignition delay period. When the induced charges are consumed, the current quickly drops to zero, causing a decrease in electric field forces and the jet to roll back, resulting in the discharge being extinguished. However, the discharge voltage differs from that measured in classical EDM [24]. During the E-Jet discharge process, a very sluggish voltage rising process has emerged. This is mostly due to the back-off period following the discharge of the E-Jet tip. In this process, the surface tension forces outweigh the electric field forces, leading to retreating of the jet. Some induced undischarged charges remain on the jet surface during this period; nonetheless, the charges on the E-Jet surface do not increase throughout this whole retreating phase. Simultaneously, the opposite remaining induced charges of the same magnitude are still on both sides of the workpiece, gradually approaching balance. As a result, the voltage between the solid electrode and the workpiece decreases gradually to zero in the meantime. Moreover, the time it takes for the voltage to slowly rise between the E-Jet and the workpiece is the same as the time it spends for the voltage to slowly fall between the solid electrode and the workpiece.

3.2.2. Discharge Sequence Analysis

Two types and eight groups of experiments were designed to capture the waveform of a single discharge and count the sequence of discharges in order to understand the relationship between the discharges on both sides. the gap between the E-Jet nozzle and the sacrifice workpiece surface was fixed at first, and the distance between the solid electrode and the workpiece randomly adjusted. The discharge signals were sampled at four-level distances. In the second group, the distance between the solid electrode and the workpiece was fixed, and the distance between the nozzle and the sacrifice workpiece randomly regulated. All discharge signals are collected and a histogram was created, as shown in Figure 5. The abscissa represents the leading time difference of the E-Jet electrode EDM and the solid electrode EDM.

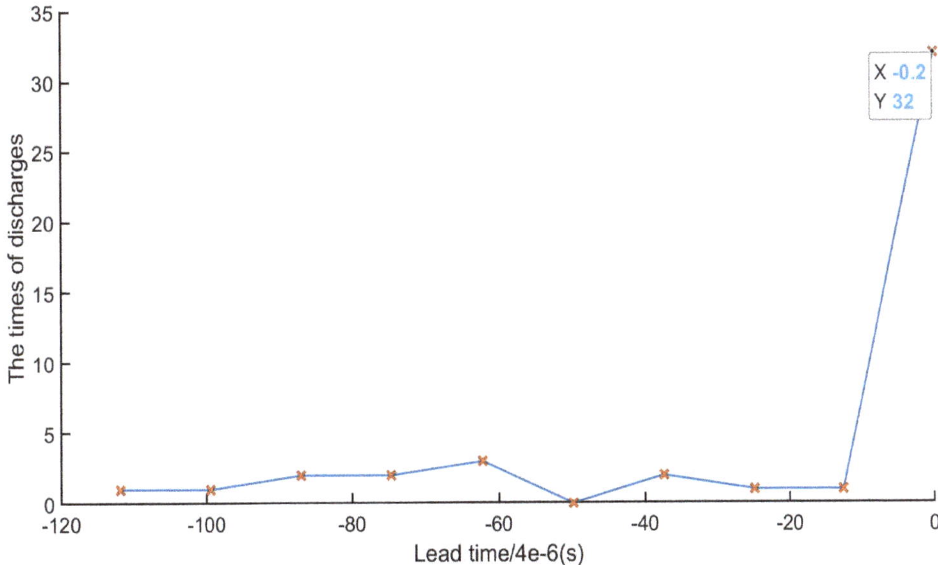

Figure 5. Time difference of discharge process between E-Jet EDM and solid electrode EDM.

The discharge at both ends occurred almost simultaneously during the experiment. However, according to the sampling data analyzed by Matlab software, as shown in Figure 5, the E-Jet electrode EDM is mostly ahead of 1–2 discharge sample units before the solid electrode EDM, that is, about $(4–8) \times 10^{-6}$ s. This occurrence sequence is independent of the control mode, regardless of whether the solid electrode or the E-Jet nozzle is adjusted. This reveals that the E-Jet EDM process induces the discharge of solid electrode. Moreover, if retaining the distance between the E-Jet and the workpiece, adjusting the position of the solid electrode relative to the workpiece can result in discharge. Furthermore, if fixing the distance between the solid electrode and the workpiece, adjusting the gap between the E-Jet nozzle and the workpiece can also result in discharge. As a result, this method supports two types of gap servo control.

3.3. Influence of Distance on Discharge Energy

The gap servo control is the primary selected method in the traditional EDM to ensure stable continuous processing. As a result, the influence of the distance on the discharge energy should be considered. Because of the high-frequency noises in the high-voltage discharge process, it is difficult to calculate the discharge energy accurately and directly. We use a windowed filter to deal with high-frequency signals before calculating the discharge energy. Because the discharge energy is roughly equal to the heat generated during the discharge process [25], it can be expressed as follows:

$$E_e = \int_0^{t_e} u_e(t) \cdot i_e(t) \cdot dt \qquad (1)$$

where U_e denotes the discharge voltage, I_e denotes the discharge current, and t_e denotes the discharge duration. If described more precisely, this formula can be rewritten as:

$$E_e = \sum_{0}^{t_e/\Delta t} u_e(t) \cdot i_e(t) \cdot \Delta t \tag{2}$$

where Δt is the sampling interval.

3.3.1. Influence of the Gap on Solid Electrode Side on Discharge energy

At first, the distance between the E-Jet nozzle and the sacrificial workpiece was set to 0.5 mm and the distance between the solid electrode and the workpiece surface to 0.195 mm. The concentration of the electrolyte (NaCl) was 5% by weight. Then, 2.8 kV was applied between the E-Jet electrode and the solid electrode. All other machining parameters were maintained as constant, and the distance between the solid electrode and workpiece was changed progressively by manually adjusting the 3D platform (B) knob in Figure 2 from 0.195 mm to 0.18 mm at 0.005 mm intervals and prohibiting the discharge from arcing or breaking off. The EDM discharge energy on the solid tool electrode side was calculated by recording the voltage and current at various discharge sites. The sacrificial workpiece and the workpiece in the experiment were both silicon, allowing the machining features to be visible to the naked eye.

When the distance between the jet nozzle and the workpiece stayed constant, as illustrated in Figure 6a, the discharge duration time decreased with increasing distance between the solid electrode and the workpiece. The discharge current dropped as the distance between the electrodes increased, as illustrated in Figure 6b, ranging from 0.1 A at 0.18 mm to 0.04 A at 0.195 mm. The gap voltage between the solid electrode and the workpiece steadily grew at a low level throughout the discharge process while the gap was less than 0.185 mm, but it quickly increased from 0.185 to 0.19 mm, from dozens of volts to 230 V, as shown in Figure 6c.

The discharge energy may be computed indirectly using Equation (2). The discharge energy between the solid electrode and the workpiece, as well as the discharge energy between the E-Jet and the workpiece, decreased steadily as the gap distance between the solid electrode and the workpiece increased, as shown in Figure 7. On the jet side, the discharge energy was always larger than that on the solid electrode side. The narrower the gap distance between the solid electrode and the workpiece, the bigger the discharge energy difference between the jet electrode side and the solid side.

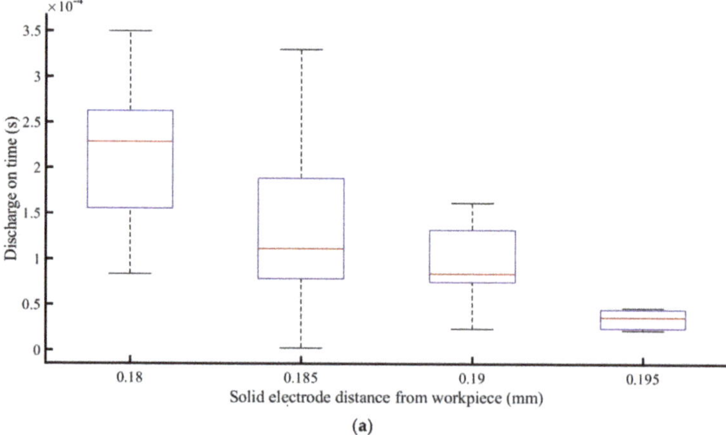

(a)

Figure 6. *Cont.*

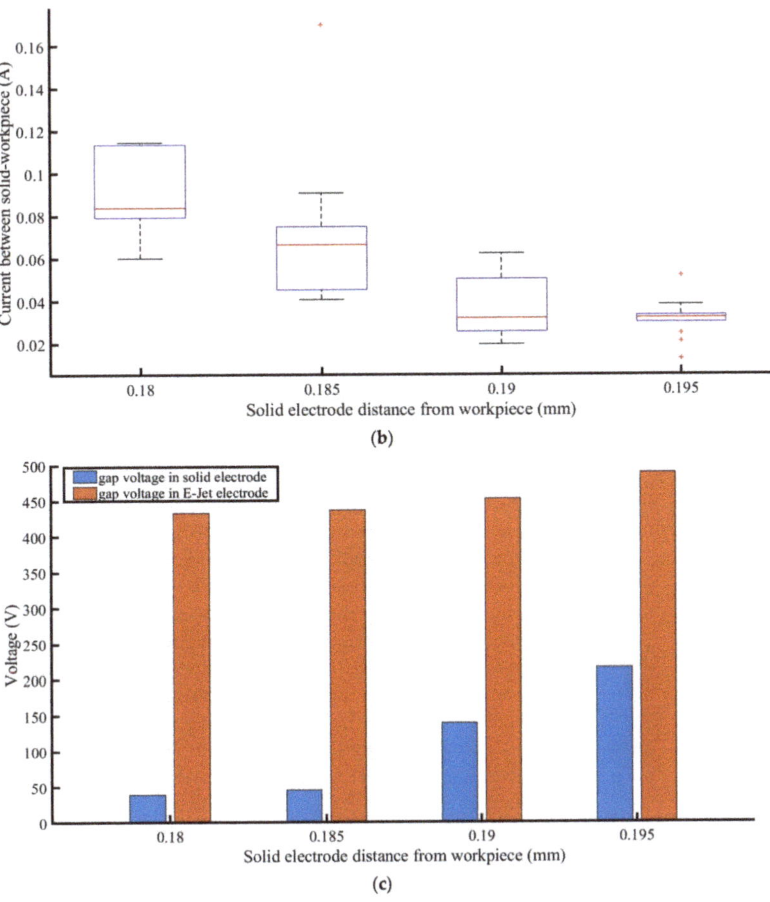

Figure 6. Analysis of the discharge duration (**a**), gap current (**b**), and gap voltage (**c**) changes with the variation of the distance between solid electrodes and workpiece.

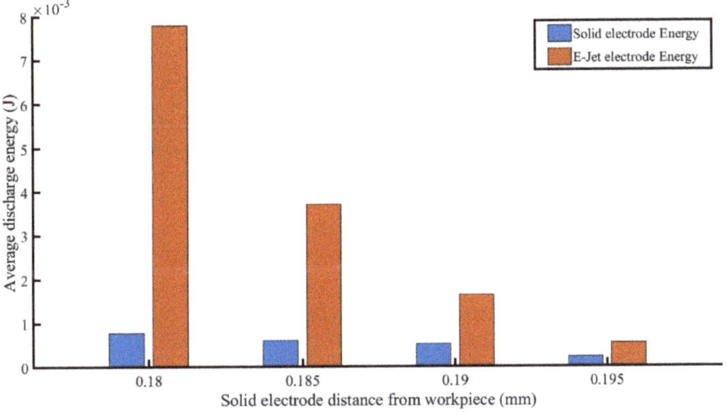

Figure 7. The discharge energy changes with the variation of the distance between solid electrodes and workpiece.

These might be caused by the reasons that in the discharge process, the capacity between the solid electrode and the workpiece can be treated as the feed capacity to that between the jet end and the sacrificial electrode, which in turn impacts the discharge between the solid electrode and workpiece. Changing the distance between the solid electrode and the workpiece entails changing the voltages between the jet end and the workpiece, as well as between the workpiece and the solid electrode. When the distance between the solid electrode and the workpiece is reduced, the capacitance between the solid electrode and the workpiece rises. Although the capacitance between the jet nozzle and the workpiece remains constant, the total capacitance between the jet nozzle and the solid electrode increases. At this stage, the strength of the electric field rises. The charge density on the cone surface increases as the electric field intensity increases, resulting in an increase in electrostatic repulsion forces at the cone tip, causing the cone angle to expand. In addition, the cone at the nozzle outlet begins to recede, resulting in a shorter cone. Meanwhile, the increased repulsive force causes the jet to expand in diameter. Heikkilä et al. [26] claimed that reducing the tip-to-collector distance can alter the form of the cone/jet in the nanofiber spinning process by the changes of the electric field. The electrolyte jet, on the other hand, can be regarded as a conductor in the E-Jet EDM process, allowing for discharge rather than continuous spinning. The higher the intensity charge density, the longer the discharge duration (Figure 6a) and the higher the current (Figure 6b). As the distance grows, the surface charge density falls gradually, the electrostatic repulsion on the cone's surface lowers gradually, the cone's length increases gradually, the cone angle decreases gradually, and the initial diameter of the jet decreases gradually. A drop in surface charge density and a small diameter of the jet result in a very short duration, and a minor charge depletion leads to a decrease in electric field forces and results in jet retraction. Because the interelectrode voltage is directly connected to the plasma characteristics during the discharge process, the greater the maintenance voltage, the larger plasma discharge gap [27]. Simultaneously, lowering the charge density and jet diameter decreases the strength of the discharge between the jet and the workpiece, causing the voltage to progressively rise as the electrode moves away from the workpiece, as shown in Figure 6c. Furthermore, when the discharge distance of the solid electrode jet electrode increases, the surface induced charge density of the two electrodes progressively decreases, which explains why the discharge energy reduces with increasing distance.

As shown above, when the distance between the jet electrode and workpiece is fixed, the gap servo control of classical EDM may be modified in real-time to create steady EDM. At the moment, the pulsating jet electrode serves as a pulse power switch for the solid electrode discharge.

3.3.2. Influence of E-Jet EDM Gap on Discharge Energy

In the experiment, fixing the distance between the solid electrode and the workpiece, the influence of E-Jet discharge on the discharge energy of solid electrode EDM was explored by adjusting the distance between the jet nozzle and the workpiece. The electrolyte content was 5% by weight. The voltage was 2.8 kV. Gradually, the distance between the jet nozzle and the workpiece was reduced from 0.69 mm to the workpiece surface at 0.02 mm intervals. The knob of the motion platform (A) was adjusted to achieve 0.69 mm, 0.67 mm, 0.65 mm, and 0.63 mm, as illustrated in Figure 1. the changes in voltages and currents at various discharge sites were recorded, ensuring that both ends can discharge. It was explored how the distance between the jet nozzle and the workpiece influences the current, voltage, discharge duration, and energy distribution between the solid electrode and the workpiece.

The graph in Figure 8a demonstrates that the duration of the jet discharge gradually reduced as the distance between the jet end and the workpiece surface rose, meaning that the discharge pulse width decreased as the distance increased. At the same time, as seen in Figure 8b, the discharge current steadily diminished. The voltages between the E-Jet nozzle and the workpiece gradually rose as the distance between the E-Jet nozzle and the workpiece grew, as illustrated in Figure 8c. The voltage on the solid side is much lower than the voltage on the E-Jet side.

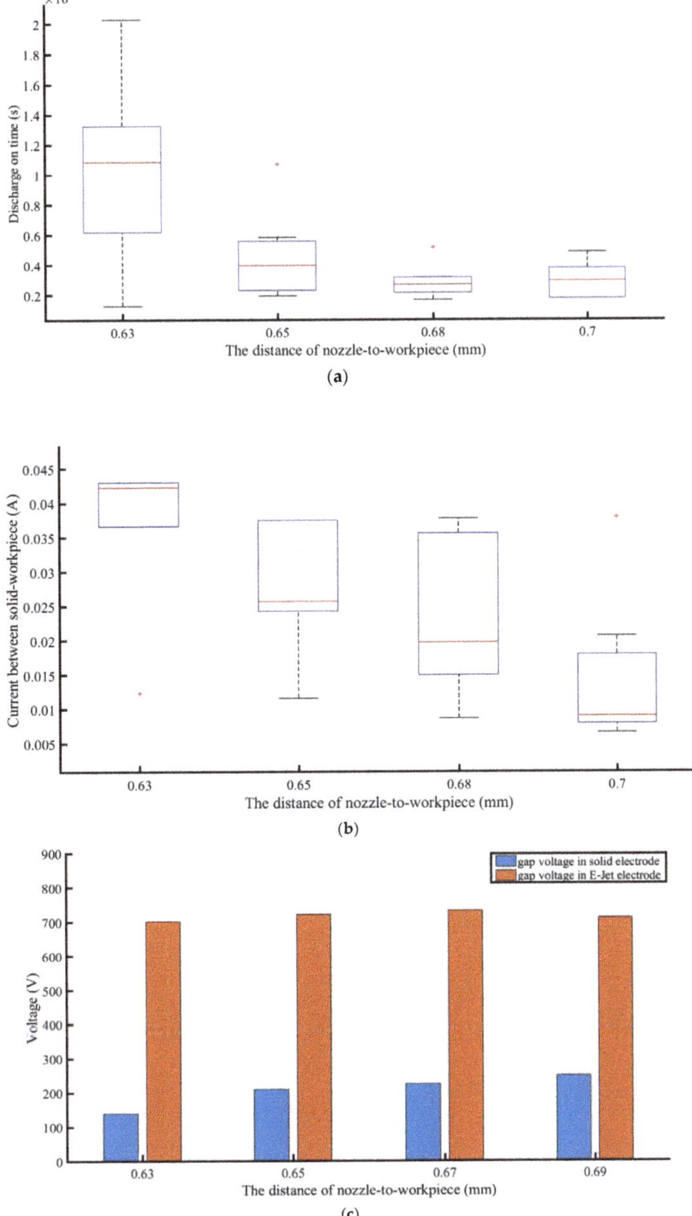

Figure 8. Analysis of the discharge duration (**a**), gap current (**b**), and gap voltage (**c**) change with the variation of the distance between E-Jet nozzle and workpiece.

The discharge energy on these two sides may be estimated using Equation (1). It is revealed that the discharge energy on the E-Jet side reduces as the distance between the nozzle and the workpiece increases. Meanwhile, as seen in Figure 9, the discharge energy on the solid side declined extremely slowly.

In this process, the jet end and the workpiece were utilized as a feed capacitor when the distance between the solid electrode and the workpiece was maintained constant and the distance between the jet nozzle and the workpiece was adjusted to discharge. The jet nozzle and the workpiece can be treated as a variable capacitor at this point, whereas the solid electrode end and the workpiece were a fixed capacitor.

The field intensity in between steadily increased as the distance between the jet nozzle and the workpiece decreased. At this point, the jet's surface charge density rose, causing the jet cone angle to rise, the cone length to fall, and the initial jet diameter to rise. Increasing charges and jet width caused a strong discharge, which increased discharge duration (Figure 8a) and discharge current (Figure 8b). This procedure is similar to that when changing the solid electrode to initiate the discharge shown in Figure 6a,b. As the distance grew, the surface charge density fell, resulting in a decrease in discharge energy, as shown in Figure 9. As the distance decreased, the larger E-jet diameter and the intensive charge density resulted in the violent plasma, gradually leading to the decrease of the maintaining voltage between the solid electrode and workpiece and the E-jet tip and the workpiece, as shown in Figure 8c.

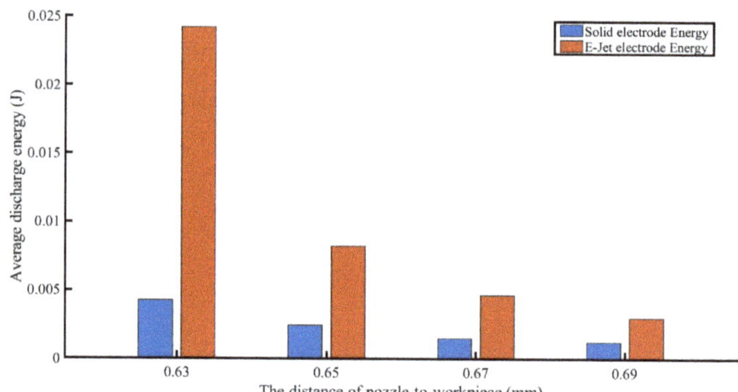

Figure 9. The discharge energy change with the variation of the distance between E-Jet nozzle and workpiece.

It can be seen from these two control approaches that the discharge between the solid electrode and the workpiece may be controlled independently of the relative distance between the jet end and the solid electrode or the distance between the solid end and the workpiece.

3.4. Machining Experiments

Experiments using single-point and groove machining were carried out to confirm the processing capacity of this technology. The solid electrode was a tiny tungsten electrode machined by electro-chemical etching method proposed by Kang and Tang [28], and the E-Jet electrolyte was a 5 wt.% NaCl aqueous solution. Polished silicon (Guangzhou Fangdao Semiconductor Co., Ltd., Guangzhou, China) was chosen as the workpiece due to its semiconducting properties, and the mirror surface allowed the machined crater and groove to be roughly observed by the naked eye before using a high magnification microscope, which is suitable for initial machining capability verification experiments.

The distance between the nozzle that generates the E-Jet and the sacrificial workpiece during the discharge process was 0.62 mm, and the distance between the solid electrode and the workpiece was 0.18 mm. The applied voltage was 2.8 kV between the nozzle and the solid electrode. Table 1 lists the machining parameters. After 1 min of processing, a point feature on the workpiece surface could be found by the naked eye. The cave could be observed through a microscope (VHX-6000 microscope, KEYENCE CORPORATION, Osaka, Japan) after being ultrasonically cleaned for 20 min and dried, as shown in Figure 10. It was discovered that a pit with a diameter of 20 µm and a depth of 9 µm could be machined. On the bottom of the pit, it was possible to see a discharge erosion phenomenon. This demonstrates that the energy generation method can be used in the solid electrode discharge process and that the materials can be removed. At the same time, the pit's shape differed a little from that of the cylindrical solid electrode with a tip in conventional EDM. This is primarily due to the fact that the solid electrode did not rotate, and the gap distance reached 0.18 mm, which is significantly greater than that in traditional EDM in gas observed by Wang and Shen [29]. Furthermore, due to the large gap, plasma drift occurred during discharges, resulting in the irregular shape of the machined pit [30].

Table 1. Machining parameters.

Parameters	Value
Solid electrode	Tungsten
Electrolyte jet electrode	5 wt.% NaCl
Workpiece	Polished silicon
Voltage	2.8 kV
E-Jet gap distance	0.62 mm
EDM gap distance	0.18 mm

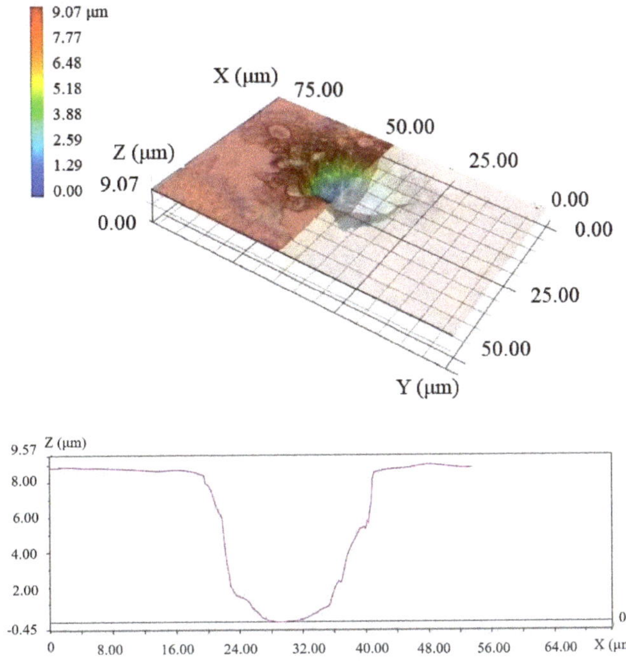

Figure 10. Analysis of the pit morphology machined by EDM in gas using solid electrode.

The machining parameters, solid electrodes, and workpieces used in the groove machining experiment were the same as that in previous pit etching process. The discharge gap between the jet electrode and the sacrificial workpiece remained constant, as did the gap between the solid electrode and the workpiece. The only difference was that the manual platform (B) under the solid electrode was slowly moved in the horizontal plane, and perpendicular to the electrode direction, to test the method's continuous groove processing feasibility. Figure 11 depicts the finished groove observed by VHX-6000 microscope. It was discovered that a regular groove with a depth of 4μm and a width of 30 μm could be processed. Although the width of the groove was fairly uniform, it was still significantly larger than the diameter of the solid electrode. This is primarily due to the plasma deflecting in a wider gap in the gas.

The experiments above show that the E-Jet EDM-based periodic discharge energy generation strategy can be decoupled and applied into the traditional EDM process. Periodic discharge can be generated automatically without the use of a traditional RC-type or transistor-type power supply. Furthermore, if the solid electrode rotates, the uncontacted E-Jet induced discharge used in traditional EDM can provide a solution for pulse power supply in micro EDM without the use of brushes.

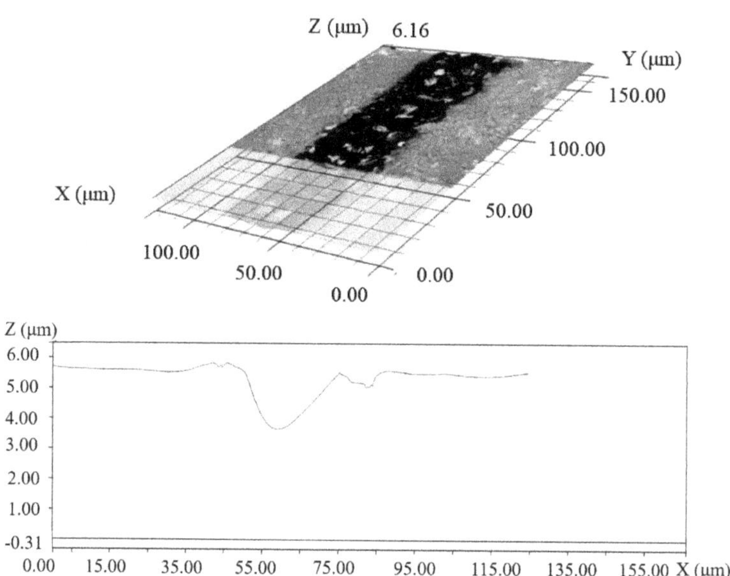

Figure 11. Continuous groove produced by EDM in gas using solid electrode.

4. Conclusions

By incorporating the E-Jet EDM energy generation method into the solid electrode EDM process, a new non-contact discharge energy generation method for traditional micro EDM was developed. This method has a wide range of potential applications in the EDM process. The study yielded the following useful findings:

1) The electrostatic induction discharge energy can be decoupled from the E-Jet EDM process to realize its application into the traditional EDM by connecting the electrical field-induced electrolyte jet EDM and the traditional EDM in series in structure.
2) It was discovered that the discharge on the E-Jet EDM side is 1–2 sampling units ahead of the discharge on the solid electrode EDM side, proving that the discharge process is induced by the E-Jet process.
3) The discharge signals analyzed demonstrated that the solid electrode EDM can be controlled by adjusting the gap between the solid electrode and the workpiece with the pulse discharge energy controlled by E-Jet EDM method. If fixing the distance between the E-Jet and the workpiece, the discharge energy per pulse will decrease with the increasing distance of the solid electrode from the workpiece.

4) The discharge energy of the solid electrode EDM can also be controlled by adjusting the gap between the E-Jet electrode and the workpiece surface, allowing the energy to be indirectly used in the solid EDM process. The discharge energy decreases with the increasing the distance between the E-Jet and the workpiece.

5) Experimental results show that this method can process a pit with a diameter of 20 μm and a depth of 9 μm and groove with a depth of 4 μm and a width of 30 μm on silicon wafers, and the results demonstrate the method's effectiveness.

The method of decoupling pulse energy from the E-Jet EDM process using two discharge devices connecting in serials proposed in this study can be applied to the energy supplying of traditional micro EDM. It can provide discharge energy by means of electric field induction without using brushes. In this way, the rotation speed of the micro electrode can be set higher, and there is no friction or contact resistance in the power supply circuit. The experimental platform will be established, and related studies will be carried out in the future.

Author Contributions: Conceptualization, Y.Z.; methodology, Y.Z., Q.G. and X.Y.; validation, Y.Z. and X.Y.; formal analysis, Y.Z. and Q.G.; investigation, X.Y.; data curation, Q.Z.; writing—original draft preparation, Y.Z.; writing—review and editing, Y.Z. and Q.G.; visualization, W.Z.; supervision, W.Z.; funding acquisition, Y.Z. All authors have read and agreed to the published version of the manuscript.

Funding: This research was funded by the National Science Foundation of China (NSFC) under grant agreement No.52175426.

Institutional Review Board Statement: Not applicable.

Informed Consent Statement: Not applicable.

Data Availability Statement: Not applicable.

Conflicts of Interest: The authors declare no conflict of interest.

References

1. Lauwers, B.; Vleugels, J.; Malek, O.; Brans, K.; Liu, K. Electrical discharge machining of composites. *Mach. Technol. Compos. Mater.* **2012**, *2012*, 202–241.
2. Xu, B.; Guo, K.; Zhu, L.K.; Wu, X.Y.; Lei, J.G.; Zhao, H.; Liang, X. The wear of foil queue microelectrode in 3D micro-EDM. *Int. J. Adv. Manuf. Technol.* **2019**, *104*, 3107–3117. [CrossRef]
3. Antar, M.; Chantzis, D.; Marimuthu, S.; Hayward, P. High speed EDM and laser drilling of aerospace alloys. *Procedia CIRP* **2016**, *42*, 526–531. [CrossRef]
4. Harane, P.P.; Wojciechowski, S.; Unune, D.R. Investigating the effect of different tool electrodes in electric discharge drilling of Waspaloy on process responses. *J. Mater. Res. Technol.* **2022**, *20*, 2542–2557. [CrossRef]
5. Casanueva, R.; Azcondo, F.J.; Alcedo, L.; Jimenez, J. Advanced Cutting Experiences for a Nuclear Power Plant Application. *IEEE Trans. Ind. Appl.* **2009**, *46*, 89–93. [CrossRef]
6. Schimmelpfennig, T.M.; Rübeling, G.; Lorenz, M. Development of novel gap control method for the Electrical Discharge Machining (EDM) of implant-supported dentures. *Procedia CIRP* **2020**, *95*, 610–614. [CrossRef]
7. He, Z.R.; Luo, S.T.; Liu, C.S.; Jie, X.H.; Lian, W.Q. Hierarchical micro/nano structure surface fabricated by electrical discharge machining for anti-fouling application. *J. Mater. Res. Technol.* **2019**, *8*, 3878–3890. [CrossRef]
8. Singh, P.; Singh, L. Examination and Optimization of Machining Parameters in Electrical Discharge Machining of UNS T30407 Steel. *J. Adv. Manuf. Syst.* **2022**, *21*, 573–589. [CrossRef]
9. Ashwin, K.J.; Muthuramalingam, T. Influence of duty factor of pulse generator in electrical discharge machining process. *Int. J. Appl. Eng. Res.* **2017**, *12*, 11397–11399.
10. Ichikawa, T.; Natsu, W. Realization of micro-EDM under ultra-small discharge energy by applying ultrasonic vibration to machining fluid. *Procedia CIRP* **2013**, *6*, 326–331. [CrossRef]
11. Chu, X.; Feng, W.; Wang, C.; Hong, Y. Analysis of mechanism based on two types of pulse generators in micro-EDM using single pulse discharge. *Int. J. Adv. Manuf. Technol.* **2017**, *89*, 3217–3230. [CrossRef]
12. Feng, G.; Yang, X.; Chi, G. Experimental and simulation study on micro hole machining in EDM with high-speed tool electrode rotation. *Int. J. Adv. Manuf. Technol.* **2019**, *101*, 367–375. [CrossRef]
13. Abbas, N.M.; Kunieda, M. Increasing discharge energy of micro-EDM with electrostatic induction feeding method through resonance in circuit. *Precis. Eng.* **2016**, *45*, 118–125. [CrossRef]
14. Koyano, T.; Kunieda, M. Achieving high accuracy and high removal rate in micro-EDM by electrostatic induction feeding method. *CIRP Ann.* **2010**, *59*, 219–222. [CrossRef]

15. Yahagi, Y.; Koyano, T.; Kunieda, M.; Yang, X. Micro drilling EDM with high rotation speed of tool electrode using the electrostatic induction feeding method. *Procedia CIRP* **2012**, *1*, 162–165. [CrossRef]

16. Zou, R.; Yu, Z.; Yan, C.; Li, J.; Liu, X.; Xu, W. Micro electrical discharge machining in nitrogen plasma jet. *Precis. Eng.* **2018**, *51*, 198–207. [CrossRef]

17. Zhang, Y.; Kang, X.; Zhao, W. The investigation of discharge restriction mechanism of the floating workpiece based electrostatic field-induced electrolyte jet (E-Jet) EDM. *J. Mater. Process. Technol.* **2017**, *247*, 134–142. [CrossRef]

18. Zhang, Y.; Han, N.; Kang, X.; Lan, S.; Zhao, W. Experimental investigation of the governing parameters of the electrostatic field–induced electrolyte jet micro electrical discharge machining on the silicon wafer. *Proc. Inst. Mech. Eng. Part B: J. Eng. Manuf.* **2018**, *232*, 2201–2209. [CrossRef]

19. Liu, W.; Luo, Z.; Kunieda, M. Electrolyte jet machining of Ti1023 titanium alloy using NaCl ethylene glycol-based electrolyte. *J. Mater. Process. Technol.* **2020**, *283*, 116731. [CrossRef]

20. Goud, M.; Sharma, A.K.; Jawalkar, C. A review on material removal mechanism in electrochemical discharge machining (ECDM) and possibilities to enhance the material removal rate. *Precis. Eng.* **2016**, *45*, 1–17. [CrossRef]

21. Singh, T.; Dvivedi, A. On performance evaluation of textured tools during micro-channeling with ECDM. *J. Manuf. Process.* **2018**, *32*, 699–713. [CrossRef]

22. Rosell-Llompart, J.; Grifoll, J.; Loscertales, I.G. Electrosprays in the cone-jet mode: From Taylor cone formation to spray development. *J. Aerosol Sci.* **2018**, *125*, 2–31. [CrossRef]

23. Kunieda, M.; Lauwers, B.; Rajurkar, K.P.; Schumacher, B.M. Advancing EDM through fundamental insight into the process. *CIRP Ann.* **2005**, *54*, 64–87. [CrossRef]

24. Yang, F.; Qian, J.; Wang, J.; Reynaerts, D. Simulation and experimental analysis of alternating-current phenomenon in micro-EDM with a RC-type generator. *J. Mater. Process. Technol.* **2018**, *255*, 865–875. [CrossRef]

25. Çakıroğlu, R.; Günay, M. Comprehensive analysis of material removal rate, tool wear and surface roughness in electrical discharge turning of L2 tool steel. *J. Mater. Res. Technol.* **2020**, *9*, 7305–7317. [CrossRef]

26. Heikkilä, P.; Söderlund, L.; Uusimäki, J.; Kettunen, L.; Harlin, A. Exploitation of electric field in controlling of nanofiber spinning process. *Polym. Eng. Sci.* **2007**, *47*, 2065–2074. [CrossRef]

27. Liu, Q.; Zhang, Q.; Zhang, M.; Yang, F. Study on the time-varying characteristics of discharge plasma in micro-electrical discharge machining. *Coatings* **2019**, *9*, 718. [CrossRef]

28. Kang, X.; Tang, W. Micro-drilling in ceramic-coated Ni-superalloy by electrochemical discharge machining. *J. Mater. Process. Technol.* **2018**, *255*, 656–664. [CrossRef]

29. Wang, X.; Shen, Y. High-speed EDM milling with in-gas and outside-liquid electrode flushing techniques. *Int. J. Adv. Manuf. Technol.* **2019**, *104*, 3191–3198. [CrossRef]

30. Dhakar, K.; Chaudhary, K.; Dvivedi, A.; Bembalge, O. An environment-friendly and sustainable machining method: Near-dry EDM. *Mater. Manuf. Process.* **2019**, *34*, 1307–1315. [CrossRef]

 **_materials_**

Article

Effect of Ultrasonic Vibration on Microstructure and Fluidity of Aluminum Alloy

An Li, Zhiming Wang * and Zhiping Sun

School of Mechanical Engineering, Qilu University of Technology (Shandong Academy of Sciences), Jinan 250353, China
* Correspondence: wzm@qlu.edu.cn

Abstract: The effect of ultrasonic vibration on the fluidity and microstructure of cast aluminum alloys (AlSi9 and AlSi18 alloys) with different solidification characteristics was investigated. The results show that ultrasonic vibration can affect the fluidity of alloys in both solidification and hydrodynamics aspects. For AlSi18 alloy without dendrite growing solidification characteristics, the microstructure is almost not influenced by ultrasonic vibration, and the influence of ultrasonic vibration on its fluidity is mainly in hydrodynamics aspects. That is, appropriate ultrasonic vibration can improve fluidity by reducing the flow resistance of the melt, but when the vibration intensity is high enough to induce turbulence in the melt, the turbulence will increase the flow resistance greatly and decrease fluidity. However, for AlSi9 alloy, which obviously has dendrite growing solidification characteristics, ultrasonic vibration can influence solidification by breaking the growing α (Al) dendrite, consequently refining the solidification microstructure. Ultrasonic vibration could then improve the fluidity of AlSi9 alloy not only from the hydrodynamics aspect but also by breaking the dendrite network in the mushy zone to decrease flow resistance.

Keywords: ultrasonic casting; aluminum alloy; fluidity; fluid simulation; dendrite growth

1. Introduction

The demand for lightweight materials in the machinery industry has led to the widespread use of aluminum alloys. However, as the structure of aluminum castings becomes increasingly complex to meet application requirements, the production of high-quality castings faces significant challenges. To address this, many efforts have been made to improve traditional casting technology. One promising method is ultrasonic-assisted casting, which can refine the grain size of alloys and improve the quality and mechanical properties of castings [1,2].

In the past, most of the ultrasonic-assisted casting process was in the form of directly inserting an ultrasonic horn into the melt [3–6]. However, due to the attenuation of ultrasonic energy, this method is not suitable for castings with complex structures and is more commonly applied in semi-continuous casting processes [7–9]. Aiming to overcome this disadvantage, attempts have been made in recent years to use an ultrasonic horn as part of the mold cavity. For example, Mukkollu et al. applied ultrasonic vibration to stainless steel molds and combined ultrasonic with slope casting to obtain ingots with finer microstructures and higher performance [10]. Peng Yin et al. used a steel clamp to fix sand molds and applied ultrasonic vibration to the resin sand mold through the clamp to refine the microstructure of aluminum casting [11]. If elaborately schemed, ultrasonic vibration applied to molds theoretically can be suitable for any complex casting, which is an important direction in realizing the application of ultrasonic vibration to complex shape castings. For castings with complex shapes and structures, the casting quality not only involves microstructure determined by solidification but also depends greatly on the casting ability related to flow, mold filling, and shrinkage [12–14]. However, previous

Citation: Li, A.; Wang, Z.; Sun, Z. Effect of Ultrasonic Vibration on Microstructure and Fluidity of Aluminum Alloy. *Materials* **2023**, *16*, 4110. https://doi.org/10.3390/ma16114110

Academic Editors: Thomas Niendorf and Jana Bidulská

Received: 13 April 2023
Revised: 12 May 2023
Accepted: 30 May 2023
Published: 31 May 2023

studies on ultrasonic-assisted casting mainly focused on its role in refining microstructure and rarely involved the effects regarding the flow and filling properties of the casting.

It has been found that wall vibration can change the velocity field of the fluid near the wall and affect the flow characteristics of the fluid [15–18]. However, excessive violent vibration can excite turbulence and consequently increase flow resistance significantly [19,20]. Ultrasonic vibration, a type of mechanical wave with a frequency exceeding 20 KHz, inevitably affects the flow field changes during the filling process of the melt. In addition, unlike pure fluids, the flow of melt is also significantly influenced by solidification factors [21,22]. In particular, ultrasonic vibration can break dendrites during solidification [23,24]. Although there are few reports, this dendrite fragmentation effect can theoretically also promote the fluidity of castings [25]. Therefore, changes in alloy fluidity under ultrasonic vibration are influenced by both solidification and hydrodynamics factors. However, there is currently limited research in this area, especially regarding the relationship between the two factors, which is not yet clear.

In order to promote the application of ultrasonic vibration in improving alloy fluidity and enhancing casting quality, in the present work, uniform ultrasonic vibration of mold walls was achieved through resonance. Based on this, the effect of ultrasonic vibration on the fluidity and solidification structure of cast aluminum alloys (AlSi9 and AlSi18 alloys) with different solidification characteristics was investigated. Furthermore, the impact of ultrasonic vibration on the flow field during the molten metal filling process was analyzed using fluid simulation software (ANSYS-FLUENT). Based on solidification structure analysis and fluid simulation, the mechanism by which ultrasonic vibration affects the fluidity of cast aluminum alloys was studied from two perspectives: the impact on dendritic growth during solidification and hydrodynamic factors.

2. Experimental Procedure

Binary hypoeutectic AlSi9 and hypereutectic AlSi18 alloys were selected to investigate the influence of ultrasonic vibration on their fluidity in the casting process. The alloys were prepared from pure aluminum (99.9 wt.%) and Al-20 wt.% with Si master alloys according to the nominal composition.

The ultrasonic vibration system used in the present work consisted of a support ultrasonic generator (TL-1200 × 3, infinitely adjustable power 0–1200 W), a 24,000 Hz ultrasonic transducer, a Ti6Al4V horn, and a metal mold (4Cr5MoSiV1), as shown in Figure 1.

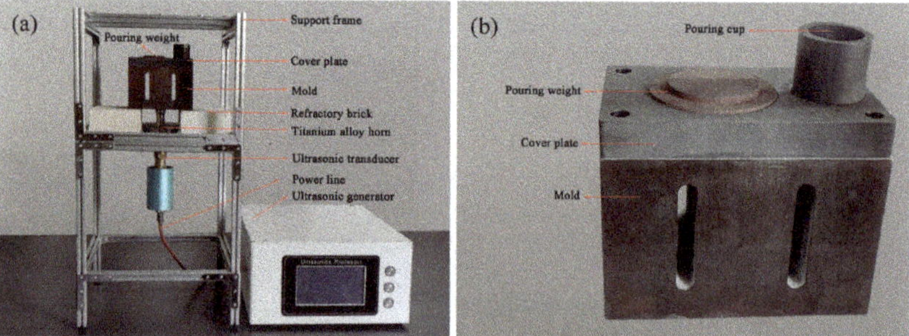

Figure 1. Ultrasonic vibration system: (**a**) experimental setup; (**b**) metal mold.

To ensure efficient vibration of the mold cavity under resonance mode, the mold's overall structure was specially designed based on modal analysis conducted using COMSOL Multiphysics software, as illustrated in Figure 2.

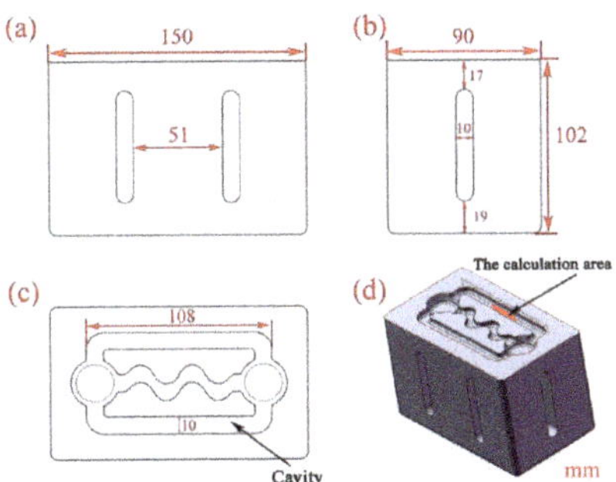

Figure 2. Fluidity test mold: (**a**) main view; (**b**) left view; (**c**) top view; (**d**) isometric view.

Figure 3 displays the vibration modal shape of the mold in the half cycle (phase shift 0–3.14) under ultrasonic excitation, as simulated using COMSOL Multiphysics software. The material properties utilized for the simulation are listed in Table 1. The results reveal that the mold vibration takes the form of elastic volume deformation, primarily along the Z axis perpendicular to the bottom of the cavity.

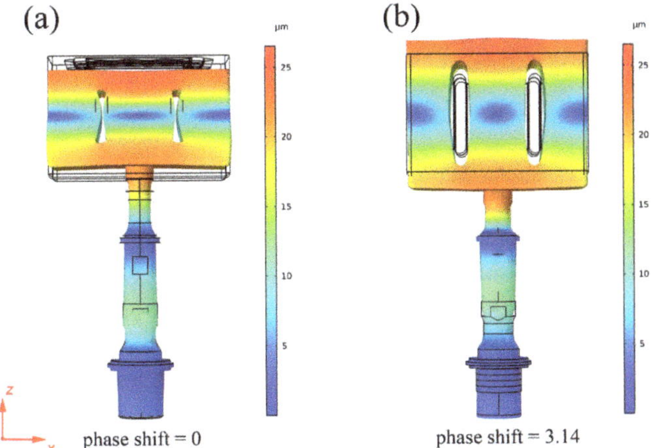

Figure 3. Simulated resonance vibration of the fluidity test mold in the half vibration cycle: (**a**) phase shift 0; (**b**) phase shift 3.14.

Table 1. Material properties used in numerical simulation.

Materials	Density (kg·m^{-3})	Young's Modulus (GPa)	Poisson's Ratio
Structural steel	7850	200	0.30
Aluminum	2700	70	0.33
Ti-6Al-4V	4510	113	0.34
H13	7000	140	0.30

Vibration of the mold cavity is the primary factor influencing the fluidity of the melt. Consequently, the displacement of the mold cavity surface was extracted and is depicted in Figure 4. The results indicate that the surface displacement of the mold cavity was relatively uniform and increased gradually with the rise in input power. The maximum surface displacement of the mold cavity (which represents the vibration intensity) can reach approximately 20 μm with an input ultrasonic power of 1080 W. The upper cover plate, which was integrated with the sprue and riser of the mold, was constructed of graphite and was not rigidly connected to the mold to avoid influencing the vibration mode, as depicted in Figure 1b.

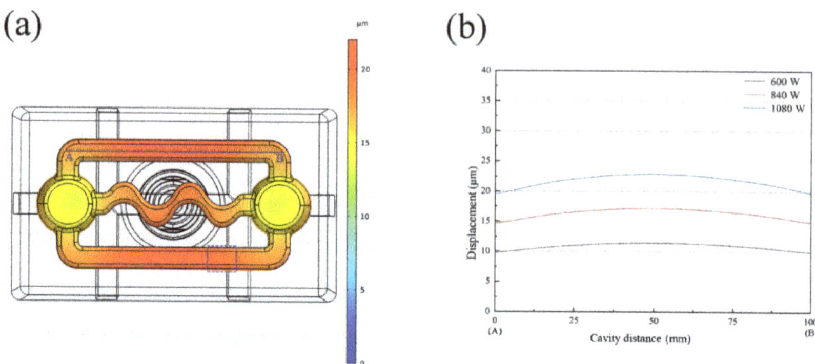

Figure 4. Vibration of mold cavity under different ultrasonic powers: (**a**) vibration mode of the mold cavity; (**b**) distribution of vibration amplitude along the blue line segment A–B in subfigure (**a**).

Aluminum alloys AlSi9 and AlSi18 were melted using a crucible resistance furnace (SG2-5-10) and subjected to slag removal and degassing before being held at 973 K and 1003 K, respectively. Prior to casting, the mold must be preheated to 473 K, connected to the ultrasonic horn, and then positioned on the bracket. The level of the mold was adjusted using a level gauge, and ultrasonic vibration was activated while pouring the aluminum alloy melt into the mold to evaluate its liquidity. After pouring, the ultrasonic vibration was turned off until solidification was complete. To assess fluidity, the cavity of the bending part in the middle of the mold was obstructed, allowing the melt to flow only through the straight sections on both sides, as shown in Figure 2. The length of the two straight sections of the fluidity sample was measured, and the average value was calculated to determine its fluidity.

To explore the impact of ultrasonic power on the fluidity and microstructure of alloys with distinct solidification modes, binary hypoeutectic AlSi9 and hypereutectic AlSi18 were chosen as the experimental materials, considering their different solidification temperature ranges. To this end, a comparative experimental plan was developed, as illustrated in Table 2.

Table 2. Schemes for the application of ultrasonic vibration.

Schemes	Alloy	Solidification Temperature Range (K)	Pouring Temperature (K)	Theoretical Degree of Superheat (K)	Ultrasonic Power (W)
1					0
2	AlSi9	865–807	973	108	600
3					840
4					1080
5					0
6	AlSi18	934–850	1003	69	600
7					840
8					1080

The analysis of ultrasonic vibration on the solidification structure was carried out on the designated position of the casting samples marked by the blue dotted box in Figure 4. After polishing, the samples were etched using Keller's reagent (2.5% HNO$_3$ + 1.5% HCL +

1% HF + 95% H_2O), and their microstructures were analyzed using Leica DM2700M microscopy (Leica, Wetzlar, Germany) and Oxford Nordy Max3 electric backscatter diffusion (EBSD) (Oxford Instruments, Abingdon, UK).

3. Results and Discussion

3.1. Effect of Ultrasonic Power on Alloy Fluidity

The impact of ultrasonic power on the fluidity of AlSi9 alloy is depicted in Figure 5. The results indicate that the fluidity of the alloy increased initially with an increase in ultrasonic power until it reached a peak and then decreased. The flow length of AlSi9 alloy substantially improved from 92.9 ± 4.5 mm to 182.0 ± 9.8 mm by applying 840 W ultrasonic vibration. However, upon further increasing the ultrasonic power to 1080 W, the flow length of the alloy started to decline, akin to the effect induced by 600 W power.

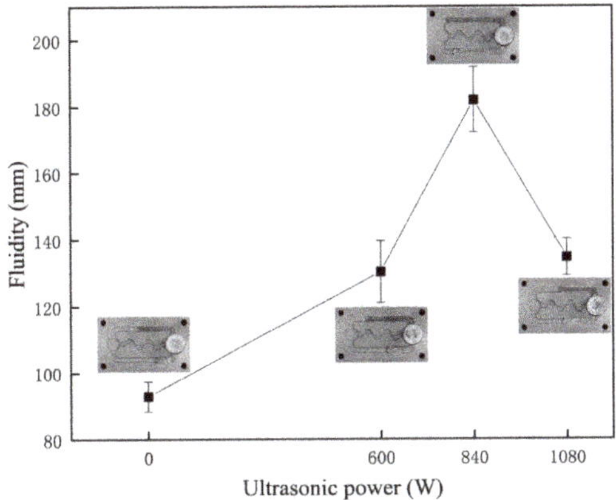

Figure 5. The influence of ultrasonic power on the fluidity of AlSi9 alloy.

In Figure 6, the effect of ultrasonic power on the fluidity of AlSi18 alloy is illustrated. The results indicate that as the ultrasonic power increased, the fluidity of AlSi18 alloy followed a similar trend of initially increasing and then decreasing. However, when compared to AlSi9 alloy, the influence of ultrasonic vibration on the fluidity of AlSi18 alloy was significantly weaker, as evidenced by a maximum flow length increase of only 26.2 ± 12.9 mm under 840 W ultrasonic vibration.

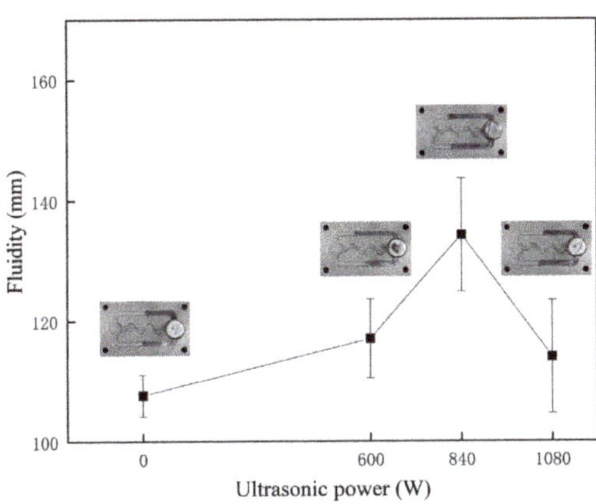

Figure 6. The influence of ultrasonic power on the fluidity of AlSi18 alloy.

3.2. Effects of Ultrasonic Vibration on the Solidification Microstructure of Alloys

Based on the above results, it is evident that the effect of ultrasonic vibration on the fluidity of AlSi9 alloy is significantly higher than that of AlSi18 alloy, which may be attributed to their distinct solidification characteristics. To further investigate the influence of ultrasonic vibration on the solidification structure of both alloys, a comparative microstructure analysis was conducted at the same position indicated by the blue dotted box in Figure 4.

As shown in Figure 7, the effect of ultrasonic vibration on the microstructure of hypoeutectic AlSi9 alloy was analyzed. In the absence of ultrasonic vibration, primary α (Al) exhibited a typical dendritic morphology, and coarse needle-like eutectic silicon was distributed between the dendrites. Upon the application of ultrasonic vibration, the α (Al) dendrites tended to break into short rods, and the degree of breaking was proportional to the ultrasonic power. Moreover, the secondary dendrite spacing at the sampling position decreased significantly with the increase in ultrasonic vibration power, as demonstrated in Table 3.

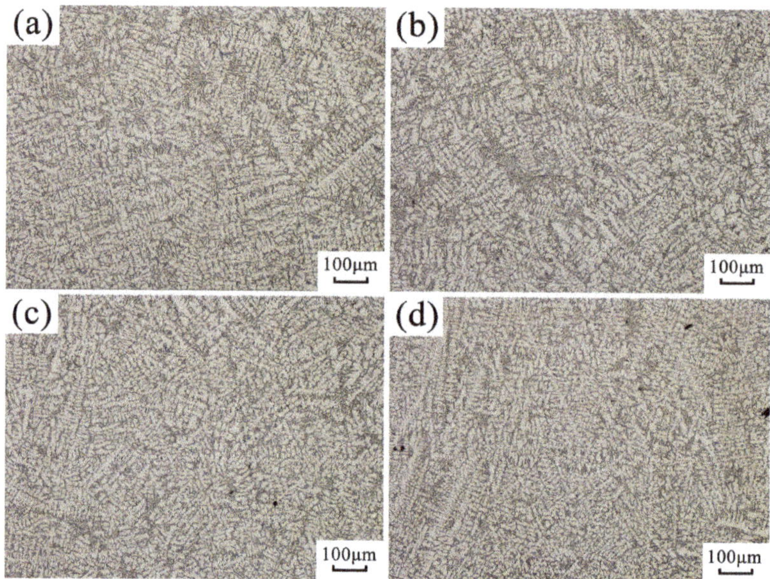

Figure 7. Effects of ultrasonic vibration on the solidification microstructure of hypoeutectic AlSi9 alloy: (**a**) no ultrasonic vibration; (**b**) ultrasonic power 600 W; (**c**) ultrasonic power 840 W; (**d**) ultrasonic power 1080 W.

Table 3. Influence of ultrasonic vibration on secondary dendrite spacing of AlSi9 alloy.

Ultrasonic Power (W)	Average Secondary Dendrite Spacing (μm)
0	12.45 $\pm$ 0.33
600	9.64 $\pm$ 0.34
840	8.42 $\pm$ 0.17
1080	7.55 $\pm$ 0.34

Different from AlSi9 hypoeutectic alloy, the primary crystalline phase of hypereutectic AlSi18 alloy is the Si phase. The effect of ultrasonic vibration on its microstructure is illustrated in Figure 8, which indicates that ultrasonic vibration had no significant impact on the morphology and size of primary silicon and the eutectic structure.

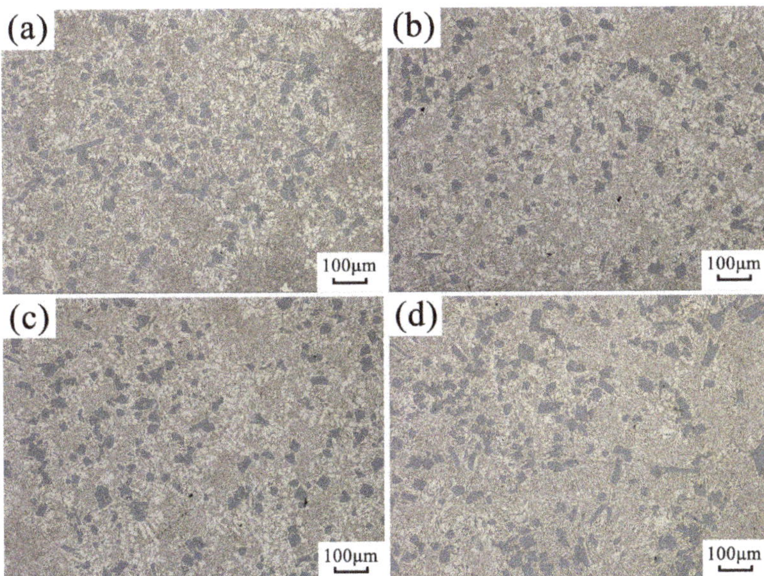

Figure 8. Effect of ultrasonic vibration on the solidification microstructure of hypoeutectic AlSi18 alloy: (**a**) no ultrasonic vibration; (**b**) ultrasonic power 600 W; (**c**) ultrasonic power 840 W; (**d**) ultrasonic power 1080 W.

Previous studies have demonstrated that ultrasonic vibration can disrupt the growth of dendrites during solidification through cavitation and acoustic flow mechanisms [23,24,26], which can ultimately enhance the fluidity of the alloy [25]. In this regard, the acoustic pressure distribution in the alloy melt was calculated, assuming that the cavity was completely filled with melt without any solidification. The density and sound velocity of liquid alloy at a temperature of 973 K were 2.35×10^3 kg·m^{-3} and 5.496×10^3 m·s^{-1}, respectively. Figure 9 illustrates that the maximum acoustic pressure in the melt ranged from 3–8 MPa at different input powers, which exceeds the reported cavitation threshold of 1 MPa in aluminum alloy [1]. It is noteworthy that all the ultrasonic power levels used in this experiment could theoretically generate cavitation effects in the melt near the mold cavity wall. However, during solidification, when the shell begins to form, the actual ultrasonic power propagates to the melt adjacent to the solidification front, and it will subsequently weaken. Therefore, it is necessary to further investigate whether cavitation can occur in the actual melt during solidification.

As discussed above in Figure 7, the dendrites of AlSi9 alloy were effectively broken by ultrasonic vibration, consequently helping to improve the fluidity of the alloy. The dendrites here may be broken by ultrasonic cavitation or ultrasonic mechanical vibration.

It has been proven that high-energy ultrasonic vibration can promote the heterogeneous nucleation of primary silicon and effectively refine the primary silicon of AlSi18 alloy [27]. However, in the present work, the size and morphology of primary silicon in AlSi18 alloy shown in Figure 8 were not influenced by ultrasonic vibration. This may indirectly verify that due to the high solidification rate and fast formation of the solidified shell, ultrasonic vibration attenuates when it passes through the interface between the solidified shell and the mold. Therefore, the ultrasonic intensity transmitted to the melt at the solidification front is not enough to produce a cavitation effect, and the solidification front is then only affected by the mechanical ultrasonic vibration and acoustic flow effect. Under this condition, the effect of ultrasonic cavitation promoting the heterogeneous nucleation of primary silicon by improving the wettability of nucleation particles [28] has not occurred here. In addition, the crystallization of primary silicon is in the form of non-dendritic and faceted growth, so the mechanical vibration and acoustic flow caused by ultrasonic vibration also cannot break the growing primary silicon. Therefore, the effect of ultrasonic vibration on refining grain by promoting heterogeneous nucleation and broken dendrites is not reflected in hypereutectic AlSi18 alloy. However, even in the absence of cavitation phenomena within the melt, forced convection generated by ultrasonication can fragment the coarse dendritic grains [5]. The hypoeutectic AlSi9 alloy used in this study also has well-developed α (Al) dendritic structures, and its dendritic structure was significantly refined under the action of forced convection (as shown in Figure 8). For AlSi9 alloy with a well-developed α (Al) dendritic structure, the continuously growing dendrites during solidification will form a net-like structure, thereby increasing flow resistance inside the melt. When subjected to ultrasonic vibration of the mold, the well-developed dendritic structure inside the melt is destroyed, reducing the flow resistance of the mushy zone, which is very beneficial for promoting the fluidity of the melt. Since hypereutectic AlSi18 alloy does not have an obvious dendritic structure during solidification, the ultrasonic vibration of the mold cannot reduce the flow resistance inside its melt by destroying dendrites, thereby improving the fluidity of the alloy. Therefore, in this study, the promoting effect of mold ultrasonic vibration on the fluidity of AlSi9 alloy was much better than that of AlSi18 alloy.

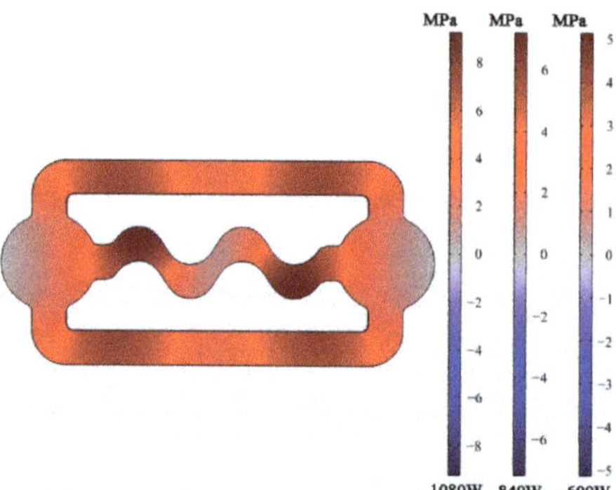

Figure 9. Simulated acoustic pressure (root mean square) in the melt.

The grain size and distribution of both AlSi9 and AlSi18 alloys were analyzed further using electron backscatter diffraction (EBSD), as presented in Figures 10 and 11. As shown in Figure 10, ultrasonic vibration significantly reduced the average grain size of AlSi9 alloy. With the increase in ultrasonic power, the average grain size decreased significantly, and the grain refinement effect was found to be proportional to the ultrasonic input power, as summarized in Table 4. Therefore, this indicates that ultrasonic vibration-induced fragmentation of the primary α (Al) dendrites of AlSi9 alloy not only enhances its fluidity but also refines the grain size through dendrite detachment and multiplication. However, for the hypereutectic AlSi18 alloy shown in Figure 11, both the primary Si and eutectic structure were not significantly refined by ultrasonic vibration, consistent with the above deduction.

Table 4. Average grain size (µm) of AlSi9 alloy.

Ultrasonic Power (W)	Average Grain Size (µm)
0	352 ± 23
840	214 ± 37
1080	153 ± 16

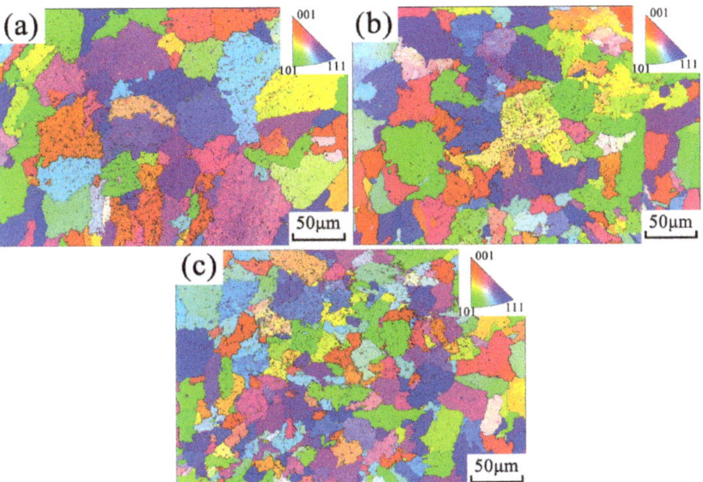

Figure 10. EBSD maps of AlSi9 alloy obtained under different power ultrasonic vibrations: (**a**) 0 W; (**b**) 840 W; (**c**) 1080 W.

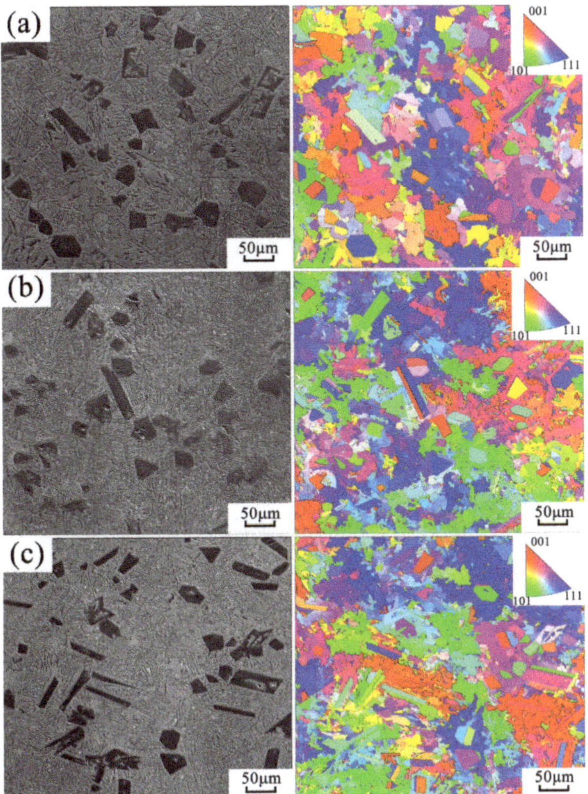

Figure 11. EBSD maps of AlSi18 alloy obtained under different power ultrasonic vibrations: (**a**) 0 W; (**b**) 840 W; (**c**) 1080 W.

As discussed above, ultrasonic vibration can effectively enhance the fluidity of the melt by breaking dendrites, especially for hypoeutectic AlSi9 alloy. However, the extent of dendrite fragmentation was proportional to the ultrasonic power (0–1080 W), and the fluidity of the alloy did not improve further at high power (1080 W) but instead decreased. Furthermore, although ultrasonic vibration did not change the solidification and corresponding structure of AlSi18, it did have an impact on its fluidity to some extent. Therefore, it is speculated that ultrasonic vibration may not only affect fluidity by interfering with the solidification process but also through the hydrodynamic aspects of melt flow, which will be discussed below.

3.3. Effect of Ultrasonic Vibration on the Hydrodynamics of Melt Flow

"ANSYS-FLUENT" was used to examine the effect of ultrasonic vibration on the flow field of aluminum alloy melt. The middle section of the straight channel of the mold cavity was extracted as the calculation area (long 50 mm × high 5 mm), as shown in Figure 12. To obtain accuracy of the computational results, it was necessary to generate boundary layer mesh during the process of meshing the model, as illustrated in Figure 13. The near-wall treatment adopted enhanced wall function. Solver selected the pressure solver. The SIMPLE algorithm was used for pressure and velocity coupling.

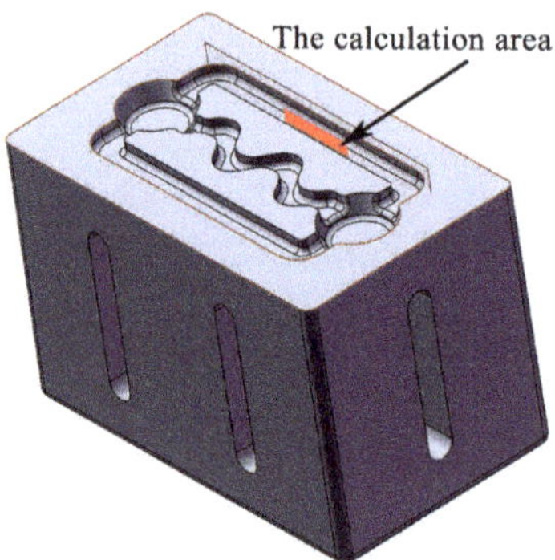

Figure 12. Calculation area for ANSYS-FLUENT.

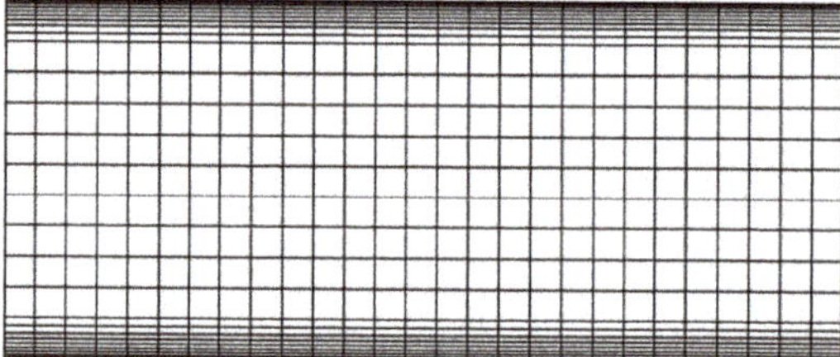

Figure 13. Schematic diagram of boundary layer mesh.

To apply ultrasonic vibration, a user-defined function (UDF) was used to define the vibration displacement and velocity boundary conditions on the bottom of the calculated region. The bottom wall was made to vibrate perpendicularly to the axis direction (x-axis direction) of the straight passage, and the harmonic vibration speed rule was defined using Formula (1). The value of vibration amplitude A was chosen based on the displacement value (ranging from 0 to 25 μm) in Figure 4, which depends on the ultrasonic power used. The time step was determined by the vibration period (T) of the wall: $T/16$.

$$v = 2\pi f A \cos(2\pi f t), \tag{1}$$

where f is frequency (24,000 Hz), t is time, and A is only related to amplitude.

The flow models used for the simulation calculation depended on the flow conditions, which could be influenced by ultrasonic vibration. Thus, in the simulation process, the laminar flow model was first used to simulate fluid flow. The flow rate of the fluid was obtained from the simulation results, and then the Reynolds number of the fluid was calculated using Formula (2). When the Reynolds number of the fluid exceeded 2300, the k-ε turbulence model was then used to simulate the fluid.

$$\mathrm{Re} = \rho v d / \mu, \tag{2}$$

where ρ is the density (10^3 kg·m^{-3}), v is the flow rate, d is the characteristic length (4.55×10^{-3} m), and μ is the viscosity coefficient (1.01×10^{-3} Pa·s).

After measurement, the flow velocity of the molten metal in the runner during casting was determined to be approximately 1.5 m·s^{-1}, and this velocity was used as the initial velocity for the fluid simulation. As shown in Table 5, the Reynolds coefficient of the fluid increased with the vibration amplitude of the wall, indicating an increasing disturbance in the flow field. When the wall vibration amplitude reached 20 μm, the Reynolds number exceeded 2300, and turbulence occurred.

Table 5. Reynolds number of fluids in different ultrasonic vibration amplitude simulations.

Amplitude (μm)	Reynolds Number
0	284
5	314
10	338
15	362
20	2634
25	3246

Figure 14 illustrates the relative velocity vector and streamline diagrams of the flow field obtained under a wall vibration amplitude of 15 μm. The results show that the upward and downward ultrasonic vibrations of the wall induce diagonal fluid flows, with periodic action occurring in one ultrasonic vibration cycle. A schematic description of the periodic action of the fluid is provided in Figure 15. During wall vibration, the angle (α) between the streamline and wall varies from $-\alpha$max to $+\alpha$max with the change in vibration phase position. Then, $+\alpha_{max}$ and $-\alpha_{max}$ are the two extremes corresponding to the $\pi/2$ and $3\pi/2$ phase positions, respectively. The periodic upward and downward flow diversion of the fluid could influence wall shear stress. In one vibration cycle, the wall shear stress increases, with α varying from 0 to $+\alpha_{max}$, and decreases, with α varying from 0 to $-\alpha_{max}$. The impact of periodically varying wall shear stress on the fluid can be measured by its flow resistance. Figure 16 displays the average flow resistance of fluid over 30 cycles under ultrasonic vibrations of different amplitudes. It was observed that the flow resistance decreased with ultrasonic vibration amplitudes of less than 15 μm. However, when the ultrasonic vibration intensity reached a certain level (amplitude > 20 μm or input ultrasonic power > 1080 W), turbulence occurred in the fluid, greatly increasing the flow resistance and leading to a loss of flowing kinetic energy.

In the flow process of AlSi9 and AlSi18 alloys under ultrasonic vibration, only from the hydrodynamics aspect, the melt flow resistance decreased with the increase in ultrasonic vibration power from 0 to 840 W, especially for AlSi18 alloy, whose fluidity under ultrasonic vibration was almost irrelevant to solidification behavior. However, when the ultrasonic power reaches 1080 W (amplitude > 20 μm), turbulence will occur in the melt, resulting in a serious loss of the flowing kinetic energy of the melt. As discussed above, the fluidity of AlSi9 alloy was improved by ultrasonic vibration not only from the hydrodynamics aspect but also greatly from the solidification aspect (breaking the growing dendrites in the mushy zone). However, the flowing energy loss caused by turbulence under 1080 W ultrasonic vibration (amplitude > 20 µm) greatly counteracted the promoting effect on the fluidity by breaking dendrites.

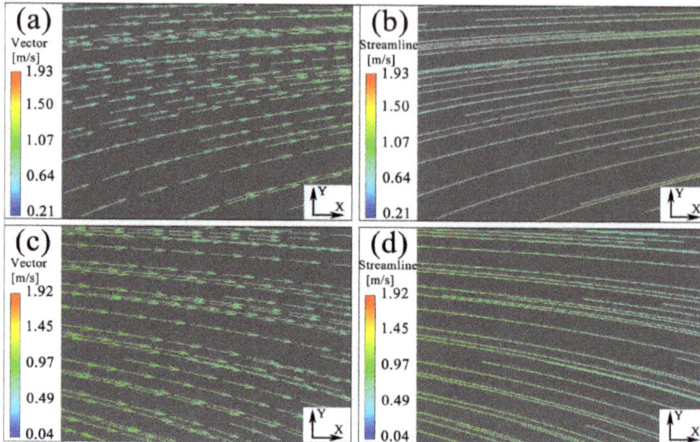

Figure 14. Velocity and streamline diagram of fluid at $\pi/2$ and $3\pi/2$ phase: (**a**) velocity diagram of fluid at $\pi/2$ phase; (**b**) streamline diagram of fluid at $\pi/2$ phase; (**c**) velocity diagram of fluid at $3\pi/2$ phase; (**d**) streamline diagram of fluid at $3\pi/2$ phase.

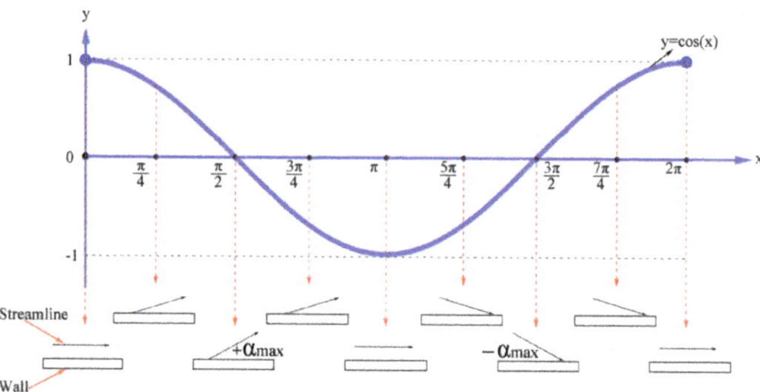

Figure 15. Phase-dependent variation of the angle between streamline and wall.

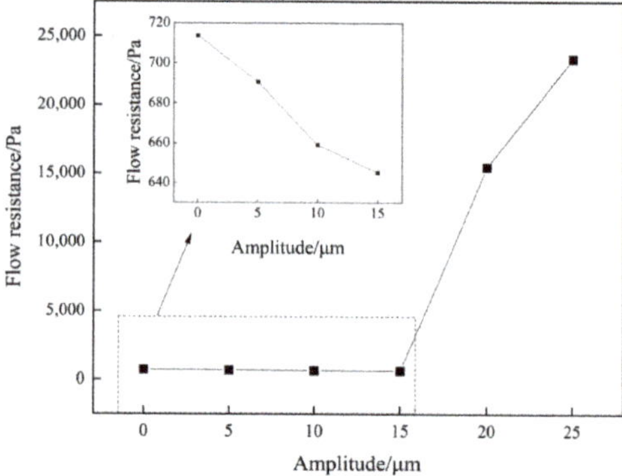

Figure 16. Variation of periodic average flow resistance with frequency under different amplitudes.

It can be found that ultrasonic vibration can affect the fluidity of alloys from the two aspects of hydrodynamics and dendrite growth. For the alloy without obvious dendrite growth characteristics (AlSi18 alloy), ultrasonic vibration affected the fluidity of the alloy mainly from the hydrodynamics aspects: ultrasonic vibration can reduce the flow resistance of melt to improve fluidity. However, high-intensity vibration (amplitude > 20 μm, ultrasonic power > 1080 W) could induce turbulence in the melt, which is unfavorable for fluidity. For the alloy with typical dendrite growth in solidification (AlSi9 alloy), it can improve the fluidity of the alloy by hydrodynamically reducing the flow resistance of the melt and breaking the growing dendrites in the mushy zone, with the latter aspect being more prominent. However, the energy loss caused by the turbulence under 1080 W ultrasonic vibration (amplitude > 20 μm) largely counteracted the promoting effect of interrupting dendrites on fluidity.

4. Conclusions

This study investigated the effects of ultrasonic vibration on the solidification microstructure and fluidity of cast aluminum alloys (AlSi9 and AlSi18) with different solidification characteristics using experimental and simulation analyses. The results show that the grain size of AlSi9 alloy, with its obvious dendritic structure, was refined under the action of ultrasonic vibration, while there was no significant change in the solidification structure of AlSi18 alloy, which does not have an obvious dendritic structure. With the increase in ultrasonic power, the fluidity of both alloys showed a trend of first increasing and then decreasing, with the effect of ultrasonic vibration on the fluidity of hypoeutectic AlSi9 alloy being more significant.

It can be found that ultrasonic vibration can affect the fluidity of alloys in both solidification and hydrodynamics aspects. For AlSi18 alloy, which lacks dendrite growing solidification characteristics, the microstructure is not significantly influenced by ultrasonic vibration, and its fluidity is mainly affected by the hydrodynamics aspect. Specifically, appropriate ultrasonic vibration can reduce the flow resistance of the melt and improve fluidity. However, when the vibration intensity is high enough to induce turbulence in the melt, the turbulence will greatly increase flow resistance and decrease fluidity. On the other hand, for AlSi9 alloy, with its obvious dendrite growing solidification characteristics, ultrasonic vibration can affect solidification by breaking the growing α (Al) dendrites, leading to refinement of the solidification microstructure. Ultrasonic vibration can then improve the fluidity of AlSi9 alloy not only from the hydrodynamics aspect but also by breaking the dendrite network in the mushy zone to decrease flow resistance, with the latter aspect being more prominent. However, the energy loss caused by turbulence under high-intensity ultrasonic vibration can largely counteract the promoting effect of breaking dendrites on fluidity.

Author Contributions: Conceptualization, Z.W. and Z.S.; experiment design, material preparation, and data collection, A.L.; writing—original draft preparation, A.L.; writing—review and editing, A.L.; formal analysis, A.L.; investigation, A.L.; supervision, Z.W. and Z.S. All authors have read and agreed to the published version of the manuscript.

Funding: This work was supported by the Natural Science Foundation of Shandong province (grant number ZR2021ME023), the Innovation Team Project of Jinan (2019GXRC035), and SQ project (2021370113124591).

Institutional Review Board Statement: Not applicable.

Informed Consent Statement: Not applicable.

Data Availability Statement: The data presented in this study are available from the corresponding author upon reasonable request.

Conflicts of Interest: The authors declare no conflict of interest.

References

1. Eskin, G.I.; Eskin, D.G. *Ultrasonic Treatment of Light Alloy Melts*, 2nd ed.; CRC Press: Boca Raton, FL, USA; Gordon and Breach: London, UK, 2015.
2. Xuan, Y.; Nastac, L. The role of ultrasonic cavitation in refining the microstructure of aluminum based nanocomposites during the solidification process. *Ultrasonics* **2018**, *83*, 94–102.
3. Barbosa, J.; Puga, H. Ultrasonic melt processing in the low pressure investment casting of Al alloys. *J. Mater. Process. Technol.* **2017**, *244*, 150–156. [CrossRef]
4. Chen, M.; Liu, Z.; Zheng, Q.; Sun, Q.; Zheng, B. Rapid preparation of B4Cp/Al composites with homogeneous interface via ul-trasound assisted casting method. *J. Alloys Compd.* **2021**, *858*, 157659. [CrossRef]
5. Zhang, L.; Li, X.; Liu, Z.; Li, R.; Jiang, R.; Guan, S.; Liu, B. Scalable ultrasonic casting of large-scale 2219AA Al alloys: Experiment and simulation. *Mater. Today Commun.* **2021**, *27*, 102329.
6. Qing-song, Y.; Gang, L.; Cheng, L.; Jia-li, S. Effect of synergistic action between ultrasonic power and solidification pressure on secondary dendrite arm spacing of vacuum counter–pressure casting aluminum alloy. *Chin. J. Nonferrous Met.* **2017**, *27*, 51–56.
7. Lebon, G.S.B.; Salloum-Abou-Jaoude, G.; Eskin, D.; Tzanakis, I.; Pericleous, K.; Jarry, P. Numerical modelling of acoustic streaming during the ultrasonic melt treatment of direct-chill (DC) casting. *Ultrason. Sonochem.* **2019**, *54*, 171–182. [CrossRef] [PubMed]
8. Zhang, L.; Li, X.; Li, R.; Jiang, R.; Zhang, L. Effects of high-intensity ultrasound on the microstructures and mechanical properties of ultra-large 2219 Al alloy ingot. *Mater. Sci. Eng. A* **2019**, *19*, 763. [CrossRef]
9. Jiang, R.; Zhao, W.; Zhang, L.; Li, X.; Guan, S. Microstructure and corrosion resistance of commercial purity aluminum sheet manufactured by continuous casting direct rolling after ultrasonic melt pre-treatment. *J. Mater. Res. Technol.* **2023**, *22*, 1522–1532. [CrossRef]
10. Mukkollu, S.R.; Kumar, A. Comparative study of slope casting technique in integration with ultrasonic mould vibration and conventional casting of aluminum alloy. *Mater. Today Proc.* **2020**, *26*, 1078–1081. [CrossRef]
11. Yin, P.; Xu, C.; Pan, Q.; Guo, C.; Jiang, X. Effect of ultrasonic field on the microstructure and mechanical properties of sand-casting AlSi7Mg0. 3 alloy. *Rev. Adv. Mater. Sci.* **2021**, *60*, 946–955. [CrossRef]
12. Bai, Y.; Mao, W.; Gao, S.; Tang, G.; Xu, J. Filling ability of semi-solid A356 aluminum alloy slurry in rheo-diecasting. *J. Univ. Sci. Technol. Beijing Miner. Metall. Mater.* **2008**, *15*, 48–52. [CrossRef]
13. Prukkanon, W.; Srisukhumbowornchai, N.; Limmaneevichitr, C. Influence of Sc modification on the fluidity of an A356 aluminum alloy. *J. Alloys Compd.* **2009**, *487*, 453–457. [CrossRef]
14. Dong, G.; Li, S.; Ma, S.; Zhang, D.; Bi, J.; Wang, J.; Starostenkov, M.D.; Xu, Z. Process optimization of A356 aluminum alloy wheel hub fabricated by low-pressure die casting with simulation and experimental coupling methods. *J. Mater. Res. Technol.* **2023**, *24*, 3118–3132. [CrossRef]
15. Jung, W.J.; Mangiavacchi, N.; Akhavan, R. Suppression of turbulence in wall-bounded flows by high-frequency spanwise oscil-lations. *Phys. Fluids A Fluid Dyn.* **1992**, *4*, 1605–1607. [CrossRef]
16. Laadhari, F.; Skandaji, L.; Morel, R. Turbulence reduction in a boundary layer by a local spanwise oscillating surface. *Phys. Fluids* **1994**, *6*, 3218–3220. [CrossRef]
17. Trujillo, S.M.; Bogard, D.G.; Ball, K.S. Turbulent boundary layer drag reduction using an oscillating wall. In Proceedings of the 4th Shear Flow Control Conference, Snowmass Village, CO, USA, 29 June–2 July 1997; p. 1870.
18. Choi, K.S.; Graham, M. Drag reduction of turbulent pipe flows by circular-wall oscillation. *Phys. Fluids* **1998**, *10*, 7–9. [CrossRef]
19. Wu, X.J.; Wang, H.G. Study on the effect of rigid plate vibration on flow field. *J. Eng. Therm. Energy Power* **2020**, *35*, 162–168.
20. Zhang, W.Y.; Yang, W.W.; Jiao, Y.H.; Zhang, D.W. Numerical study of periodical wall vibration effects on the heat transfer and fluid flow of internal turbulent flow. *Int. J. Therm. Sci.* **2022**, *173*, 107367. [CrossRef]
21. Eskin, D.G.; Tzanakis, I.; Wang, F.; Lebon, G.; Subroto, T.; Pericleous, K.; Mi, J. Fundamental studies of ultrasonic melt processing. *Ultrason. Sonochem.* **2019**, *52*, 455–467. [CrossRef]
22. Huang, H.; Qin, L.; Tang, H.; Shu, D.; Yan, W.; Sun, B.; Mi, J. Ultrasound cavitation induced nucleation in metal solidification: An analytical model and validation by real-time experiments. *Ultrason. Sonochem.* **2021**, *80*, 105832. [CrossRef]
23. Wang, S.; Kang, J.; Guo, Z.; Lee, T.; Zhang, X.; Wang, Q.; Deng, C.; Mi, J. In situ high speed imaging study and modelling of the fatigue fragmentation of dendritic structures in ultrasonic fields. *Acta Mater.* **2019**, *165*, 388–397. [CrossRef]
24. Priyadarshi, A.; Khavari, M.; Subroto, T.; Conte, M.; Prentice, P.; Pericleous, K.; Eskin, D.; Durodola, J.; Tzanakis, I. On the governing fragmentation mechanism of primary intermetallics by induced cavitation. *Ultrason. Sonochem.* **2021**, *70*, 105260. [CrossRef] [PubMed]
25. Zi-Heng, H.; Zhi-Ming, W.; Zhi-Ping, S.; Bing-Rong, Z.; Wei-Feng, R. Influence of non-uniform ultrasonic vibration on casting fluidity of liquid aluminum alloy. *China Foundry* **2022**, *19*, 380–386.
26. Wang, F.; Eskin, D.; Mi, J.; Wang, C.; Koe, B.; King, A.; Reinhard, C.; Connolley, T. A synchrotron X-radiography study of the fragmentation and refine-ment of primary intermetallic particles in an Al-35 Cu alloy induced by ultrasonic melt processing. *Acta Mater.* **2017**, *141*, 142–153. [CrossRef]

27. Feng, H.K.; Yu, S.R.; Li, Y.L.; Gong, L.Y. Effect of ultrasonic treatment on microstructures of hypereutectic Al–Si alloy. *J. Mater. Process. Technol.* **2008**, *208*, 330–335. [CrossRef]
28. Wang, F.; Eskin, D.; Mi, J.; Connolley, T.; Lindsay, J.; Mounib, M. A refining mechanism of primary Al3Ti intermetallic particles by ultrasonic treatment in the liquid state. *Acta Mater.* **2016**, *116*, 354–363. [CrossRef]

Article

Investigating the Microscopic Mechanism of Ultrasonic-Vibration-Assisted-Pressing of WC-Co Powder by Simulation

Yuhang Chen [1], Yun Wang [1], Lirong Huang [1,*], Binbin Su [2,*] and Youwen Yang [1]

[1] School of Mechanical and Electrical Engineering, Jiangxi University of Science and Technology, Ganzhou 341000, China
[2] Jiangxi Province Key Laboratory of Maglev Technology, School of Electrical Engineering and Automation, Jiangxi University of Science and Technology, Ganzhou 341000, China
*   Correspondence: huanglirong@jxust.edu.cn (L.H.); binbinsu@jxust.edu.cn (B.S.); Tel.: +86-152-1617-0316 (L.H.)

Abstract: The ultrasonic-vibration-assisted pressing process can improve the fluidity and the uneven distribution of density and particle size of WC-Co powder. However, the microscopic mechanism of ultrasonic vibration on the powder remains unclear. In this paper, WC particles with diameter 5 µm and Co particles with diameter 1.2 µm were simulated by three-dimensional spherical models with the aid of the Python secondary development. At the same time, the forming process of the powder at the mesoscale is simulated by virtue of the finite element analysis software ABAQUS. In the simulation process, the vibration amplitude was set to 1, 2, and 3 µm. Their influence on the fluidity, the filling density, and the stress distribution of WC-Co powder when the ultrasonic vibration was applied to the conventional pressing process was investigated. The simulation results show that the ultrasonic vibration amplitude has a great influence on the density of the compact. With an increase in the ultrasonic amplitude, the compact density also increases gradually, and the residual stress in the billet decreases after the compaction. From the experimental results, the size distribution of the billet is more uniform, the elastic after-effect is reduced, the dimensional instability is improved, and the density curves obtained by experimentation and simulation are within a reasonable error range.

Keywords: WC-Co powder; ultrasonic-vibration-assisted pressing process; finite element simulation; pressed billet density

Citation: Chen, Y.; Wang, Y.; Huang, L.; Su, B.; Yang, Y. Investigating the Microscopic Mechanism of Ultrasonic-Vibration-Assisted-Pressing of WC-Co Powder by Simulation. *Materials* **2023**, *16*, 5199. https://doi.org/10.3390/ma16145199

Academic Editor: Alexander Yu Churyumov

Received: 23 June 2023
Revised: 17 July 2023
Accepted: 20 July 2023
Published: 24 July 2023

1. Introduction

WC-Co-cemented carbide with high hardness, high toughness, high wear resistance, and other excellent properties is widely used in many fields, such as mining, aerospace, automobile manufacturing, oil drilling, and so on [1,2]. With the continuous improvement in social production levels, the demand for cemented carbide in the manufacturing industry is increasing. However, in the actual production process, various defects often appear in the cemented carbide billet, such as delamination, fractures, missing angles, etc. because of the poor powder fluidity and the uneven distribution of density and particle size, which greatly limits the application of its products in various fields [3]. Improving the uniformity of the billet density and particle size distribution of cemented carbide can not only improve the hardness, bending strength, fracture toughness and other comprehensive mechanical properties, but also improve the physical properties of the material itself, such as electrical conductivity, thermal conductivity, permeability, and thermal expansion coefficient, and high-density powder metallurgy materials can also result in parts having a better machining performance and machining surface [4–6].

As a result, in order to improve the process of powder metallurgy, reduce the rejection rate, and improve the comprehensive performance of products, domestic and foreign researchers have carried out much research on the movement law of powder. The research shows that the particle will produce a volume effect, surface effect, and particle convection phenomenon under the ultrasonic-vibration-assisted pressing process [7,8]. Moreover, the ultrasonic-vibration-assisted pressing process can improve the fluidity and the uneven distribution of density and particle size of WC-Co powder, which is beneficial in modeling the arrangement of particles/bulk solids. Alhazaa used vibrations in diffusion bonding and sintering during impulse-pressure-assisted diffusion bonding and found that varying pressure could indeed reduce bonding times in diffusion bonding and reduce the requirements for pre-bonding surface preparation [9]. Sliva found that containers had different cross-section shapes, but their cross-section area and weight were constant because of the optimum arrangement of spherical particles [10]. Zhao Yanbo et al. [11] studied the influence of vibration frequency and vibration amplitude on the filling effect of iron powder and obtained the best parameters for the filling effect. Zuo Miaomiao et al. [8] analyzed the differences between the internal contact force of granular medium and the force chain, porosity, and coordination number in the particle medium after the action of the ultrasonic vibration. Wang Wentao et al. [12] studied the effect of vibration on the dense packing of refractory powder and determined that the average porosity of powder first decreased and then increased with the increase in vibration time, and they pointed out that the compaction effect caused by longitudinal vibration was better than that caused by transverse vibration. Sedaghat et al. [13] proposed a physics-based constitutive model to accurately describe the deformation behavior in the process of ultrasonic-vibration-assisted forming. At the same time, the finite element method was used to conduct a numerical simulation of the upsetting and progressive forming to evaluate the accuracy of the model. From the results, by considering the dislocation dynamics and acoustic energy transfer mechanism in the material under the ultrasonic vibration, the newly established ultrasonic constitutive equation could accurately predict the acoustic–plastic behavior of the material. Meanwhile, the application of ultrasonic vibration could significantly reduce the flow stress of the material, making it become soft during the forming process. The larger the amplitude, the smaller the inflow stress. Liu Bo et al. [14] used discrete element software to simulate the influence of Nd/Fe/B permanent magnet powder on the filling density at different vibration frequencies and vibration times. The results showed that the filling density reached its maximum when the vibration frequency was between 66 and 70 Hz, and the density first increased and then stabilized with the increase in vibration time. However, there have been few studies on the pressing process of WC-Co-cemented carbide by the ultrasonic-vibration-assisted pressing process.

This work proposes the use of the ABAQUS finite element software and Python co-simulation to simulate the pressing process of WC-Co powder with the aid of ultrasonic vibration. Compared with conventional pressing, the effects of changes in the vibration amplitude on the evolution of the fluidity, filling density, and stress distribution of WC-Co powder after application of the ultrasonic vibration are explored, and the result of the pressing experiment is compared with that of the simulation.

2. Construction of the Experiment Platform

An experiment platform was constructed to apply the axial ultrasonic vibration to the traditional pressing process, as shown in Figure 1. This platform is primarily made up of five parts: an automatic press, an ultrasonic transducer, an ultrasonic generator, a mold set, and a piece of rubber cushion. The ultrasonic transducer is composed of a piezoelectric ceramic transducer, which can convert electrical energy into mechanical energy, an amplitude rod, and a tool head. The vibration frequency is 20 kHz and the driving power is 2 kW, and the ultrasonic amplitude can be steplessly adjusted from 0% to 100%. It is mounted on the automatic press with a flange whose pressure is transmitted directly to the transducer shell.

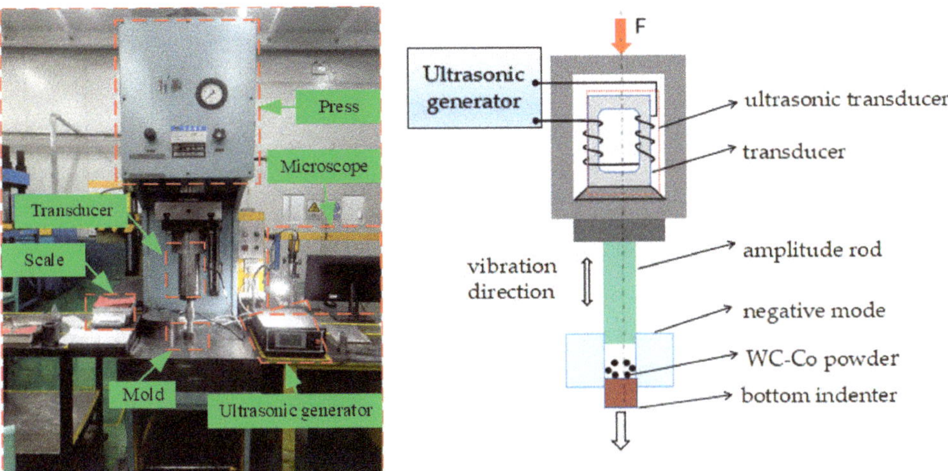

Figure 1. The experiment platform of the ultrasonic-vibration-assisted pressing process and its schematic drawing.

First, a certain mass of mixture of WC-Co powder is weighed and poured into the mold. When the mold is filled with powder, the automatic press is started to make its indenter move slowly until it comes into contact with the powder. Then, the ultrasonic transducer is started, and certain pressure is applied for a few seconds for the prepressure ultrasound. Finally, the pressure is applied slowly until the working pressure is attained. In order to calculate the density, the mass, the height and the diameter of the pressed billet are measured, and the errors of the mass and size are ±1 mg and ±0.01 mm, respectively.

3. Construction of the Mathematical Model

The continuum theory and the discrete element method are commonly used in the numerical simulation of powder materials [15]. The continuum theory has contributed to many achievements in the research on powder impact forming, which mainly regards powder as a continuum to study the dynamic mechanical response of powder in the loading process [16]. Compared with the continuum theory, the discrete element method regards powder particles as independent discrete individuals, and each discrete individual has corresponding physical properties, so the discrete element method is more in line with the actual situation. In this paper, the numerical simulation of WC-Co powder pressing process is carried out based on the discrete element method, according to Newton's second law, so the equation of motion can be obtained as shown in Equation (1):

$$M\frac{d^2X}{dt^2} + C\frac{dX}{dt} + KX(t) = f(t) \tag{1}$$

M—the mass of the particle, kg; X—the displacement of the particle, m; C—the damping coefficient; K—the elastic coefficient; t—time, s; f—the unit load, N.

The role of the ultrasonic-vibration-assisted powder pressing process is to make the powder particles inside the mold undergo violent collision contact through high-frequency vibration, resulting in greater mobility, so that some smaller particles are evenly filled into the pores, with the aim of improving the density of the billet and reducing its porosity [8]. In order to better fit the actual situation and reduce the calculation amount, the soft-ball model is used to calculate the contact force between particles. The model structure is shown in Figure 2. Particle i and particle j have contact slip under the external forces. The dotted line represents the particles just in contact, a represents the amount of overlap between the two particles in the normal direction, and b represents the tangential displacement of particle j.

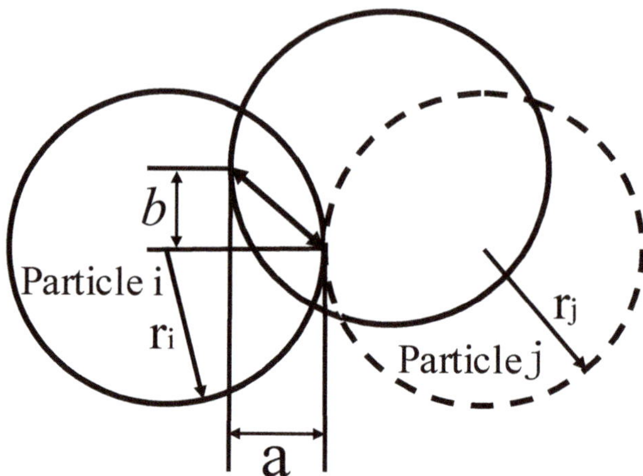

Figure 2. The contact slip between particles.

Based on the soft-ball contact mechanics model, Figure 3 is a three-dimensional simplified model, which mainly introduces damping, spring, and friction coefficients between the two particles [17], and its normal force (F_{nij}) is shown in Equation (2).

$$F_{nij} = -n[K_n a^{3/2} + C_n(V_{ni} - V_{nj})n] \qquad (2)$$

n—the unit vector of particle i and particle j; a—the normal overlap of two particles; $a = r_i + r_j - g_{ij}$; g_{ij}—the distance between two particle centers; V_{ni} and V_{nj}—the normal velocity of particle motion; K_n—the normal elastic coefficient; C_n—the normal damping coefficient. The unit vector (n) of particle i and particle j, the normal elastic coefficient (K_n), and the normal damping coefficient (C_n) are shown in Equations (3)–(5).

$$n = \frac{r_i - r_j}{|r_i - r_j|} \qquad (3)$$

$$K_n = \frac{4}{3} E^* (R^*)^{1/2} \qquad (4)$$

$$C_n = 2(m K_n)^{1/2} \qquad (5)$$

m—the mass of the particle; E^*—the effective modulus of elasticity; R^*—the effective radius of the particle; E^* and R^* can be obtained from Equations (6) and (7).

$$E^* = \frac{E_i E_j}{E_i(1 - v_i^2) + E_j(1 - v_j^2)} \qquad (6)$$

$$R^* = \frac{r_i r_j}{r_i + r_j} \qquad (7)$$

where E_i and E_j are the elastic modulus of particle i and particle j, respectively, and v_i and v_j are the Poisson's ratios of particle i and particle j, respectively.

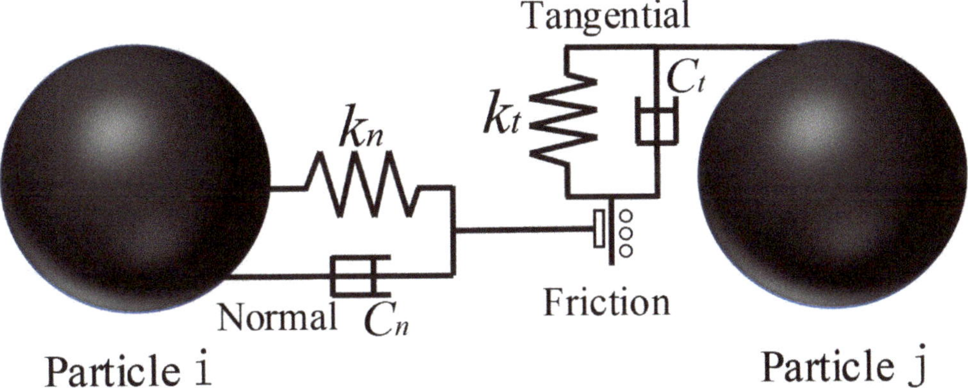

Figure 3. The contact mechanics of the soft-ball model.

The tangential force (F_{tij}) between particles is shown in Equation (8):

$$F_{tij} = -(K_t b + C_t V_t) \tag{8}$$

where b is the tangential displacement; V_t is the particle contact point velocity; K_t and C_t are the tangential elastic coefficient and the damping coefficient, respectively. K_t and C_t can be obtained from Equations (9) and (10):

$$K_t = 8a^{1/2}G^*(R^*)^{1/2} \tag{9}$$

$$C_t = 2(mK_t)^{1/2} \tag{10}$$

where G^* is the effective shear modulus, and its calculation formula is shown in Equation (11):

$$G^* = \frac{G_i(2 - v_j) + G_j(2 - v_i)}{G_i G_j} \tag{11}$$

where G_i and G_j are the shear modulus of particle i and the particle j, respectively.

4. Random Generation of Three-Dimensional Powder Particles

In this paper, finite element simulation is carried out on the microscale particles to compare the mechanism of the particle rearrangement, deformation, and interaction under the assisted pressing process of ultrasonic vibration. The entire simulation process is shown in Figure 4a. First, the electron microscopic images of WC with a particle size of 5 μm and Co powder particles with a particle size of 1.2 μm are obtained, as shown in Figure 4b,c. It can be seen from the figures that WC and Co particles are similar to spherical particles. Therefore, a three-dimensional spherical model of two particle sizes is adopted to simulate the WC and Co particles.

The random generation of powder particles is performed through the ABAQUS and Python secondary development software. The main principles are as follows:

① To set the powder particle drop area, for example, randomly drop particles with radius R in a cube with length, width, and height of 10, as shown in Figure 5. Their spherical coordinates (x,y,z) must satisfy $R \leq x \leq 10 - R$, $R \leq y \leq 10 - R$, and $R \leq z \leq 10 - R$.

② To determine whether the randomly generated balls overlap:

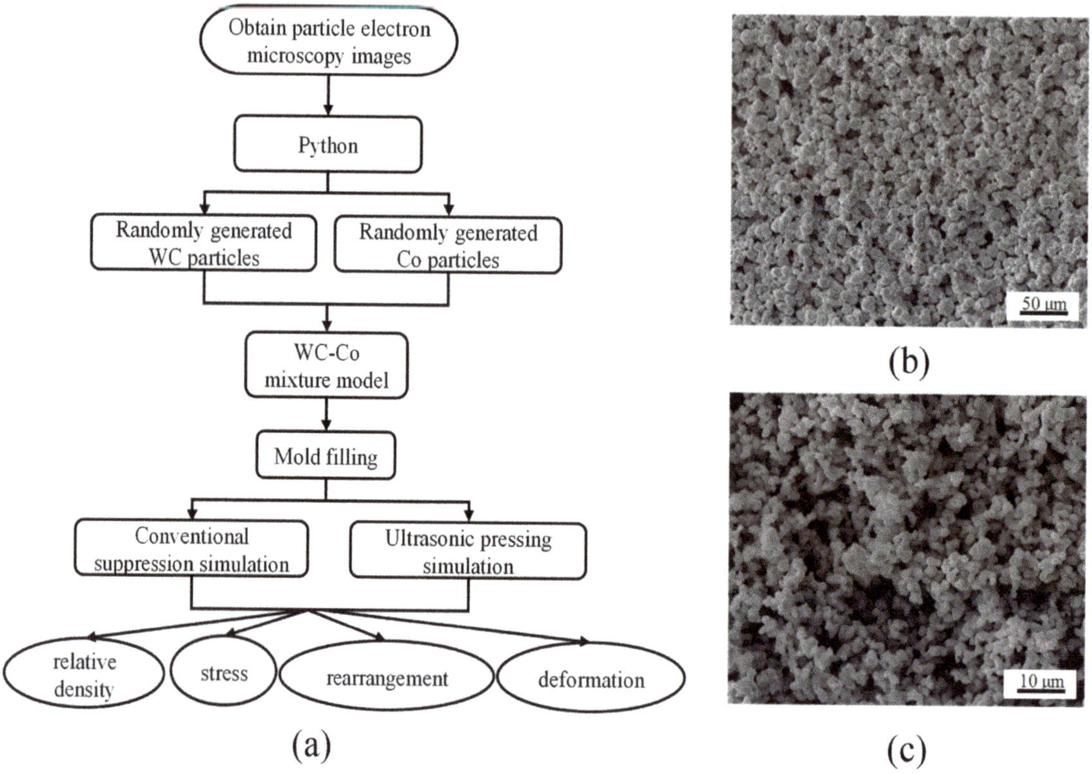

Figure 4. (a) The simulation flow chart. (b) The electron microscopic diagram of raw WC powder. (c) The electron microscopic image of raw Co powder.

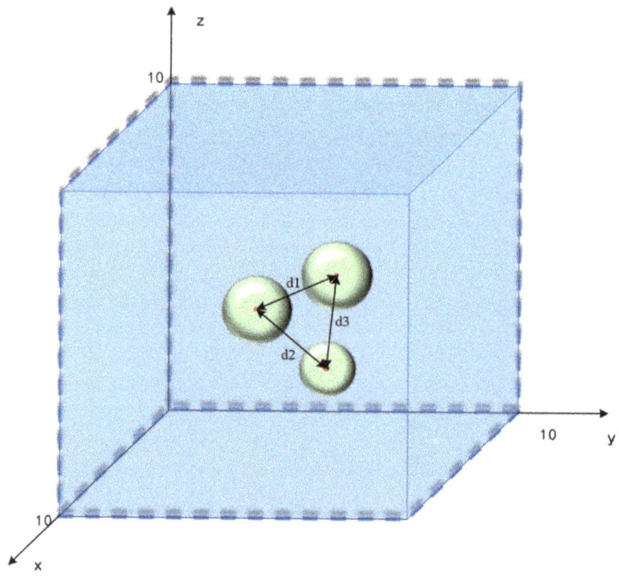

Figure 5. The schematic diagram of the particle drop.

Since two spherical particles of different particle sizes are generated, the radius of the two particle sizes is set to R_1 and R_2, respectively. The distance between the two spheres can be obtained by Formula (12):

$$d = \sqrt{(x_i - x_j)^2 + (y_i - y_j)^2 + (z_i - z_j)^2} \tag{12}$$

The distance between any two particles should meet the following conditions, as shown in Equations (13) and (14):

$$d_1 \geq 2R_1 \tag{13}$$

$$d_2 \geq R_1 + R_2 \tag{14}$$

③ The number and density of particles generated must be set and controlled.

5. Simulation Model and Parameter Setting

5.1. Material Attributes

The assignment of material properties is one of the essential steps of finite element simulation, and the ABAQUS software has a dedicated material library containing most of the commonly used materials, so the user can call or enter parameters directly. At the same time, the matching of parameter units is extremely important; the main reason is that the ABAQUS software has no fixed unit, so the user has to choose the corresponding matching unit for each quantity, and the unit of the final calculation result corresponds to the unit used. In this paper, the material properties of the corresponding parts are assigned as shown in Table 1.

Table 1. The material properties of each component.

Performance Parameter	ABAQUS Unit	Mold	WC	Co
Density	Tonne/mm^3	7.89×10^{-9}	1.56×10^{-9}	7.9×10^{-10}
Modulus of elasticity	MPa	2.09×10^5	7.14×10^5	2.09×10^5
Yield strength	MPa	no	2380	279
Poisson's ratio	no	0.269	0.19	0.3

5.2. Analysis Step Setting and Meshing

This work mainly studies the pressing process of WC-Co powder. The changes between powder particles are relatively complex, and there are multi-directional motions, rotations, collisions, contact deformations, etc. in the moving process, so this process is a highly nonlinear problem. In order to improve the efficiency of finite element calculation and save the simulation time, this paper adopts the dynamic-display analysis step to carry out the finite element simulation of the conventional pressing process and the ultrasonic-assisted pressing process.

In view of the complex particle movement in the powder pressing process, the mesh may be distorted, so the 4-node linear tetrahedral element mesh C3D4 in the dynamic-display analysis step is used in this paper. Since the parts have to be repeated many times when performing the material assignment and meshes in ABAQUS, the Python script is called to automate the material assignment and meshes, as shown in Figure 6.

5.3. Contact Properties and Boundary Conditions

When the contact properties are defined, the powder particles are deformable, while when the die and punch are set to rigid, compaction begins with the upper punch, and the die and lower punch are fixed. In the powder pressing process, the pressure is not all due to the deformation of the powder particles; part of it is consumed by the friction between the particles, the particles, and the mold wall, so that the pressure gradually decreases from top to bottom.

Here, under conventional pressing conditions, the friction coefficient of WC-Co is 0.2. Under ultrasonic pressing conditions, due to the anti-friction effect of ultrasound, the friction coefficient will be reduced by about 50%, according to Siegert's research [18–20], so the friction coefficient is 0.1.

This work adopts axial unidirectional pressing and applies uniform downward displacement and sinusoidal vibration to the upper punch for a downward displacement of 24 μm and a pressing time 1×10^{-2} s, which causes displacement time curves to be coupled. At the same ultrasonic frequency of 20 kHz, the amplitudes of 1, 2, and 3 μm were applied, respectively. The effects of different amplitudes on the fluidity, filling density, and stress distribution of the WC-Co powder were investigated. To speed up the simulation, the mold and powder filling ranges were scaled, while other physical quantities were adjusted accordingly [21].

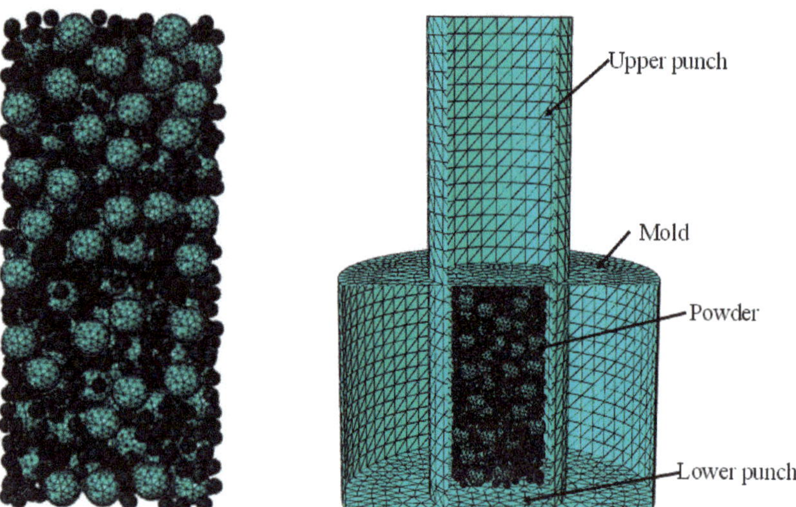

Figure 6. The WC-Co mixture and the press-model meshing.

6. Results

6.1. Analysis of Particle Flow under the Conventional and the Ultrasonic Vibration-Assisted Compression Processes

The fluidity characteristics of WC and Co particles are shown in Figure 7 under conventional and ultrasound-assisted compression conditions. In order to observe the particle fluidity more directly, a semi-sectional view is adopted. It can be seen from the figure that, without considering gravity, there are large pores between WC and Co particles in Figure 7a,e at the initial stage of compaction, and the contact force between particles is zero. When the upper punch moves down, the WC and Co particles near the upper punch first move to fill the pores, so contact slip occurs between the particles. Compared with conventional compaction, the particles move violently under the ultrasonic vibration, the particle fluidity is significantly enhanced, the arch bridge effect between particles is destroyed, and the fine particles quickly fill the pores between large particles in a short time to increase the density, and the uniformity of the particle size distribution is also improved [21,22]. As the upper punch continues to move down, the compact becomes more and more dense, and the particles gradually change from the original collision and sliding flow to mutual extrusion deformation. At this time, the increase in the compact density is mainly due to the plastic deformation of the particles. At the end of the pressing process, the height of the compact is significantly lower than that of the conventional pressing process under the action of ultrasonic vibration.

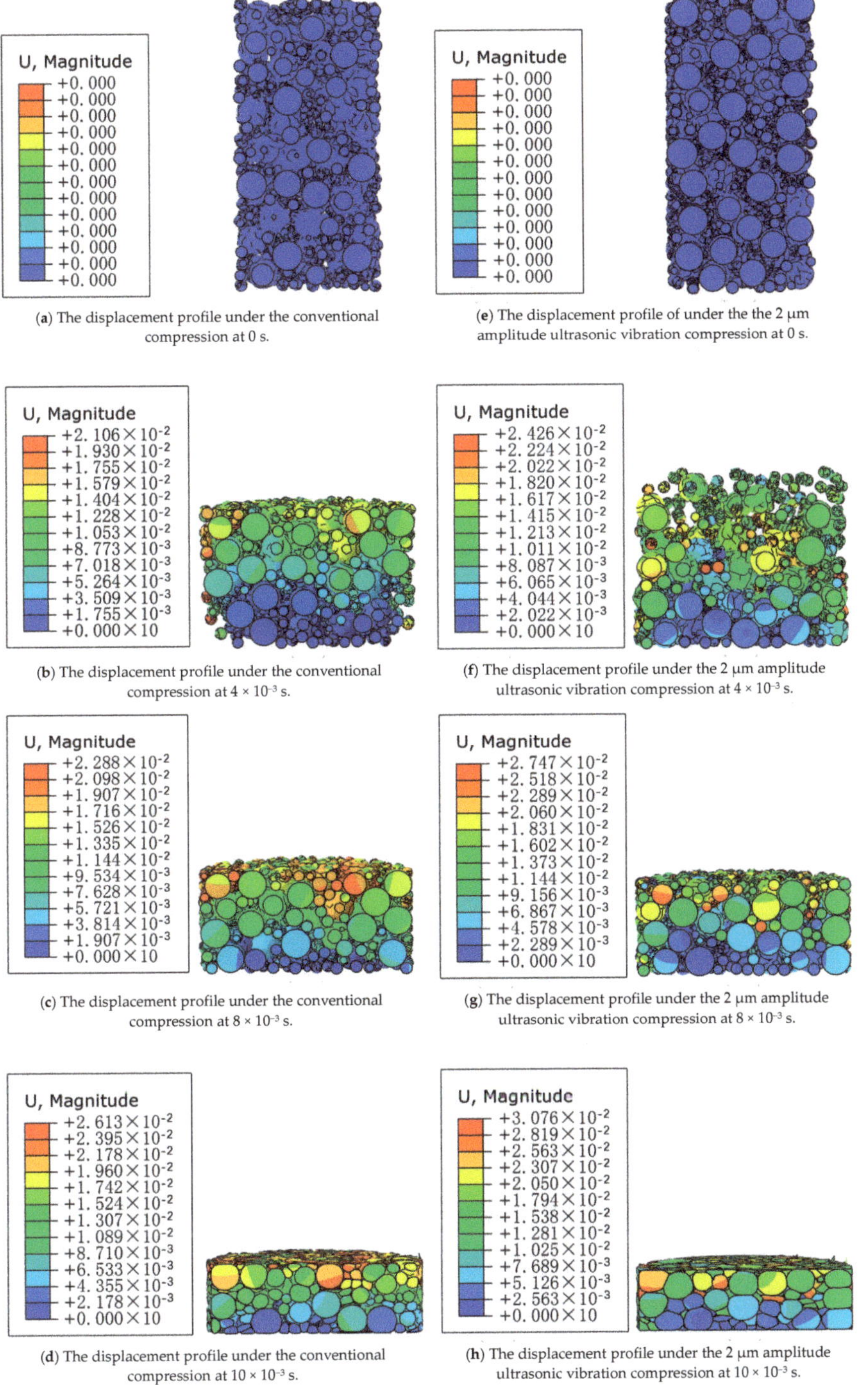

(**a**) The displacement profile under the conventional compression at 0 s.

(**e**) The displacement profile of under the the 2 μm amplitude ultrasonic vibration compression at 0 s.

(**b**) The displacement profile under the conventional compression at 4×10^{-3} s.

(**f**) The displacement profile under the 2 μm amplitude ultrasonic vibration compression at 4×10^{-3} s.

(**c**) The displacement profile under the conventional compression at 8×10^{-3} s.

(**g**) The displacement profile under the 2 μm amplitude ultrasonic vibration compression at 8×10^{-3} s.

(**d**) The displacement profile under the conventional compression at 10×10^{-3} s.

(**h**) The displacement profile under the 2 μm amplitude ultrasonic vibration compression at 10×10^{-3} s.

Figure 7. Displacement distribution cloud images at different times of the conventional pressing and ultrasonic pressing processes.

6.2. Effects of Ultrasonic Vibration Amplitudes on the Compact Density

The relationship curves between the ultrasonic amplitude and the compact density are obtained by comparison with the conventional pressing process under ultrasonic amplitudes of 1, 2, and 3 μm, respectively, as shown in Figure 8. As can be seen from Figure 8, since the initial density increase is related to the particle displacement, the compact density increases slowly. When the pressure increases to a certain extent, the compact density increases rapidly under the combined action of the plastic deformation and the displacement of the powder particles [23]. After applying different ultrasonic amplitudes, the displacement between particles is accelerated, and smaller particles fill the pores; thus, the compact density is significantly increased.

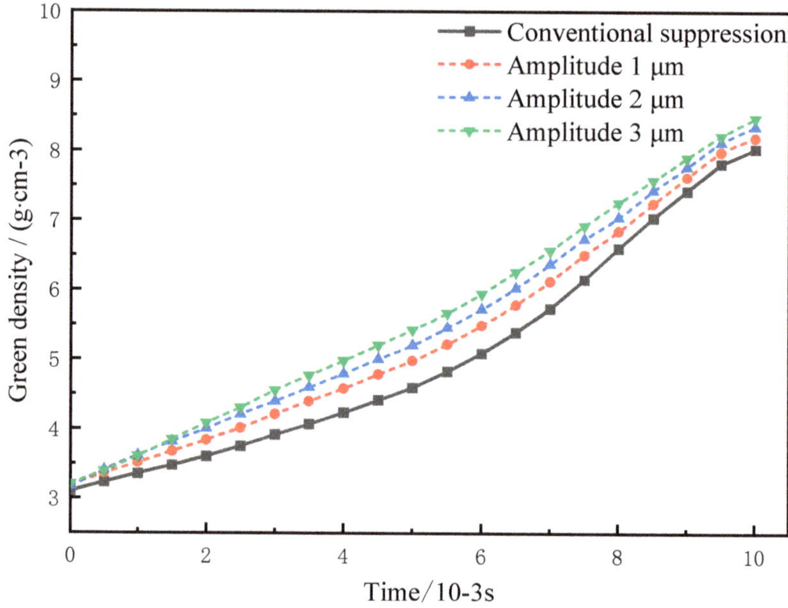

Figure 8. Effects of ultrasonic vibration amplitudes on the compact density.

6.3. Effects of the Ultrasonic Vibration on the Stress Distribution of the Compact

The Mises stress distribution cloud of WC and Co particles is shown in Figure 9 under conventional compression and different ultrasonic amplitudes. It can be seen from the figure that the powder particles are deformed after pressing, because the contact between the particles will cause stress concentration, and the WC particles are harder, so the stress concentration is more obvious. However, after the ultrasonic vibration is applied, the stress at the contact between particles gradually decreases with the increase in ultrasonic amplitude, which indicates that ultrasonic vibration can change the stress distribution between particles and reduce the deformation stress between particles. Therefore, ultrasonic vibration can reduce the stress concentration between particles during the powder pressing process, thereby reducing the residual stress in the compact after the powder pressing, reducing the elastic after-effect [24], and improving the quality of the compact.

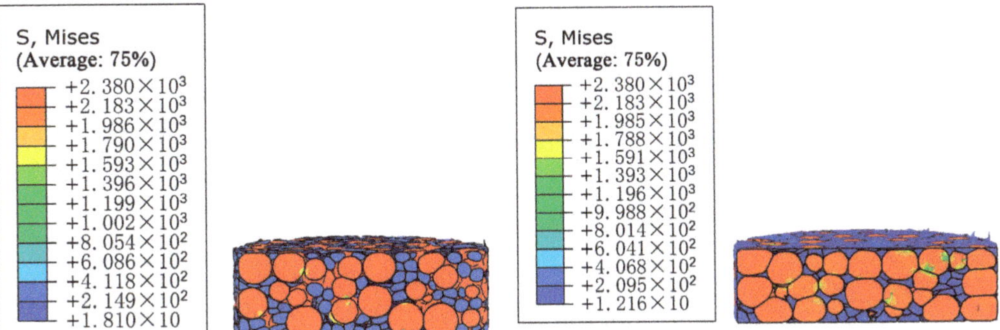

(**a**) The stress distribution under conventional pressing.   (**b**) The stress distribution under an ultrasonic amplitude of 1 μm.

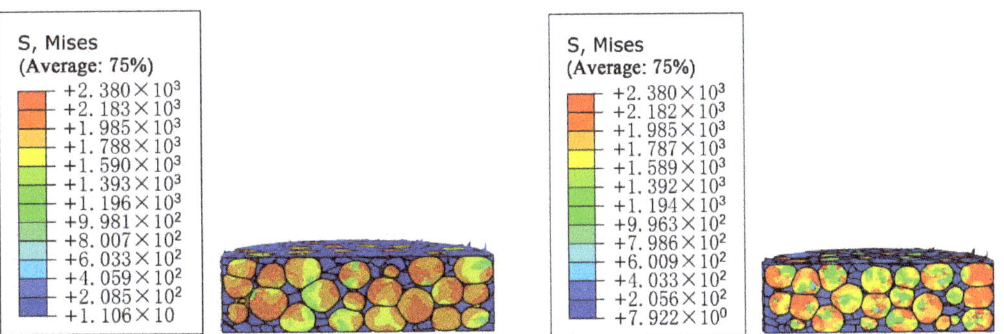

(**c**) The stress distribution under an ultrasonic amplitude of 2 μm.   (**d**) The stress distribution under an ultrasonic amplitude of 3 μm.

Figure 9. Mises stress distribution cloud images of WC and Co particles under conventional compression and different ultrasonic amplitudes.

6.4. Experimental Verification

Experiments were carried out using two kinds of pressing processes to obtain the relationship between the pressing time and the compact density, and the experimental results were compared with the simulation results when the amplitude of vibration was 3 μm, as shown in Figure 10. It can be seen from the figure that the experimental and simulated compact density errors are not large, and the errors in the forming process are both lower than 10%, indicating that the finite element simulation and experimental results are relatively consistent [25]. When the density is greater than 5 g/cm^3, the error between the experimental value and the simulation value is greater than 6%, mainly because the finite element simulation regards the friction coefficient between the particles and between the particles and the mold as a fixed value during the particle-forming process, while the friction coefficient between the particles and between the particles and the mold in the actual pressing process is a constantly changing process. The deformation and displacement of the particles in the middle and late pressing period are especially complicated, and it is difficult for a simulation to accurately reflect the forming process.

The cross-sectional microstructure of the compact under the two pressing conditions was observed and it was found that, as shown in Figure 11, the WC and Co particles were deformed correspondingly, and the particles changed from the original approximately circular state to a flat state, which is more consistent with the simulation effect. By comparing the cross sections of conventional and ultrasonic pressing compacts, it is found that the particle size distribution is more uniform under the ultrasonic pressing process, which is related to the acceleration of particle flow rearrangement under the ultrasonic action [26].

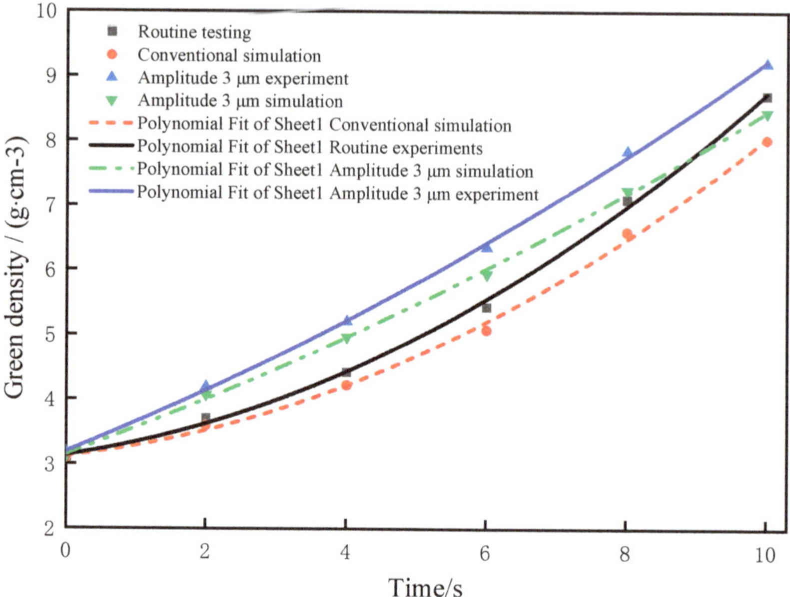

Figure 10. Results of the compression experiment and the simulation.

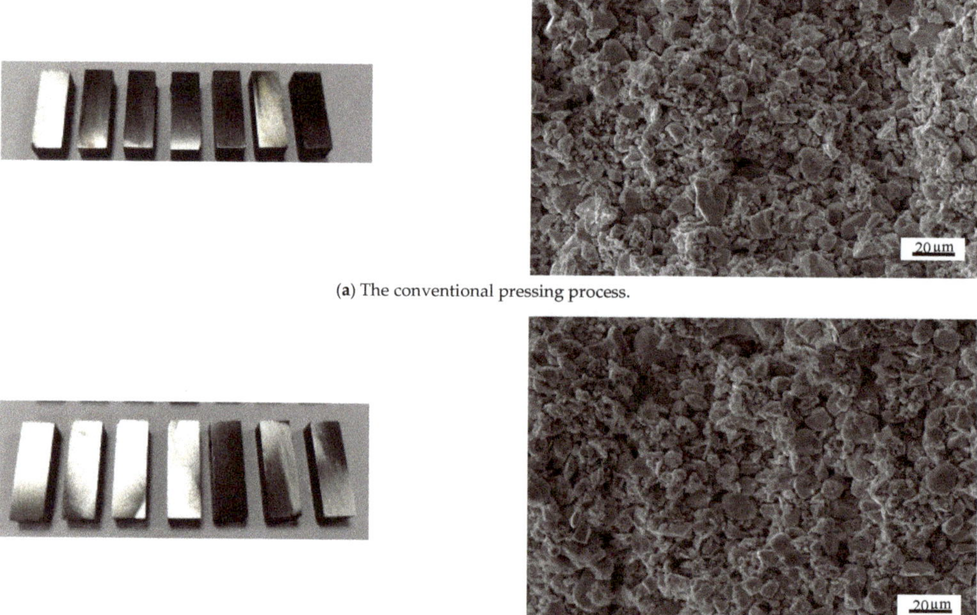

(**a**) The conventional pressing process.

(**b**) The pressing process under ultrasonic vibration action.

Figure 11. SEM of the longitudinal section of the compacts.

When the pressure is withdrawn, the powder particles will slowly return to their original state, and the compact will rebound and expand along the direction of the pressing force during and after demolding, and the compact size will increase. In order to characterize the size change after demolding, Chuxuan Chen [27] expressed as a percentage the increase in the size of the compact after demolding, as shown in Equation (15).

$$\delta = \frac{\Delta H_\mathrm{P}}{H_\mathrm{P}} \tag{15}$$

H_p—the pressed size; ΔH_p—the increased size after demolding.

Under the same pressing force for the conventional and ultrasonic vibration pressing processes, the dimensions of the compact after demolding and the dimensions of the compact after standing for 15 min and 30 min were counted. Three different positions on the compact surface were measured for each dimension, and then the average value was taken to obtain the elastic after-effects of the compact after conventional pressing and ultrasonic-vibration-assisted pressing processes, as shown in Figure 12. The δ change, as shown in Figure 13, shows that the blank size increased rapidly for about 15 min after demolding. The results show that the elastic after-effect of the compact is reduced, and the dimensional instability is slightly improved, which is consistent with the simulation results in Figure 9, indicating that the residual internal stress is reduced due to the application of ultrasonic vibration during the pressing process [13].

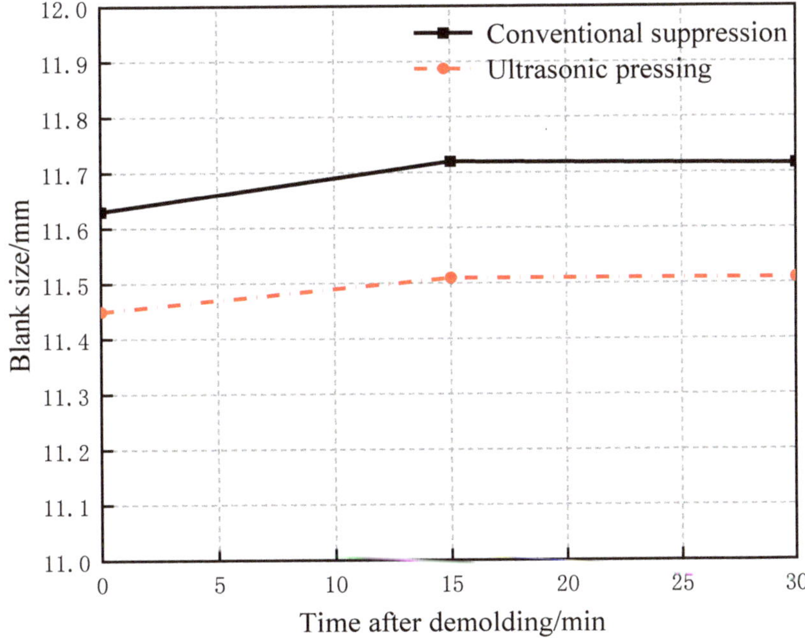

Figure 12. Changes in blank size after demolding.

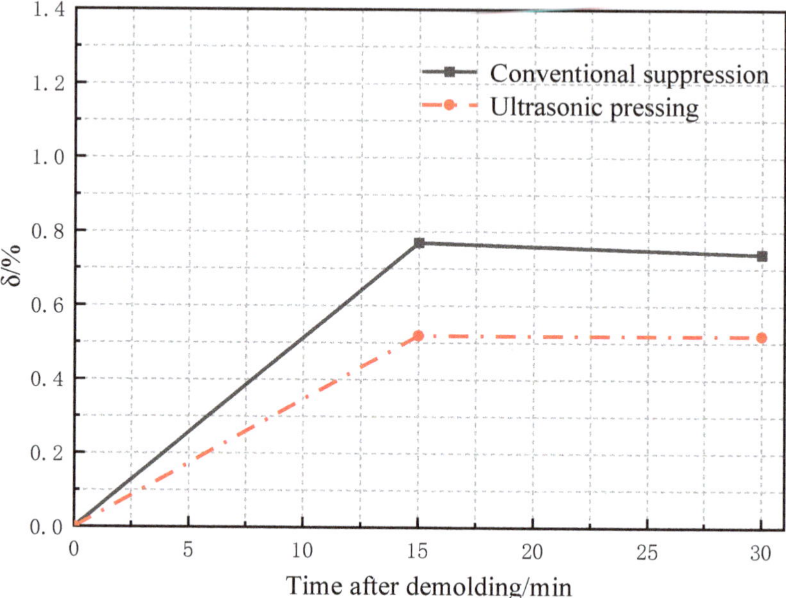

Figure 13. Changes of δ after demolding.

7. Conclusions

In this paper, Python was used to randomly generate WC and Co particles, which were automatically utilized for material assignment and meshing to reduce the modeling time. Then, ABAQUS finite element simulation and experiments were conducted on the two pressing processes respectively, and the conclusions are as follows:

(1) The influence of ultrasonic vibration amplitude on the compact density is great, so to obtain the theoretical results, values of vibration amplitudes were set 1, 2, and 3 μm to verify that the compact density increases gradually with the increase in ultrasonic amplitude. Especially in the early stage of particle deformation, the fluidity between particles is relatively intense, and the particles quickly fill the pores, so the filling density of the powder is significantly increased compared with that under conventional pressing conditions.

(2) The use of ultrasonic vibration in powder pressing can effectively reduce the deformation stress between particles, reduce the residual stress in the compact after pressing, reduce the elastic after-effect, and improve the quality of the compact.

(3) The obtained experimental results verify the developed theoretical model of the pressing process. When the value of the ultrasonic vibration amplitude is 3 μm, the finite element simulation is consistent with the experimental results, but when the density is greater than 5 g/cm^3, the error between the experimental value and the simulation value is greater than 6%, mainly because the finite element simulation regards the friction coefficient between particles and between particles and the mold as a constant value during the particle-forming process, while the friction coefficient between particles and between particles and the mold constantly changes during the actual pressing process.

Author Contributions: Conceptualization, L.H.; experiment design, material preparation, and data collection, Y.W.; writing—original draft preparation, Y.C.; writing—review and editing, Y.C.; formal analysis, B.S. and Y.Y.; investigation, Y.Y.; supervision, L.H. and B.S. All authors have read and agreed to the published version of the manuscript.

Funding: This research was funded by the Natural Science Foundation of Jiangxi province (grant number 20171BAB206030), Key Resource and Development Program for Industrial Fields, and the Innovative Leadership Program of Ganzhou Project (Kefa [2020] No. 60).

Institutional Review Board Statement: Not applicable.

Informed Consent Statement: Not applicable.

Data Availability Statement: The data presented in this study are available from the corresponding author upon reasonable request.

Acknowledgments: The authors would like to thank Zhangyuan Tungsten for conducting the pressing process experiments and the SEM inspection.

Conflicts of Interest: The authors declare no conflict of interest.

References

1. Zhang, H.; Xiong, J.; Guo, Z.X.; Yang, T.E.; Yi, J.S.; Yang, S.D.; Liang, L. Influence of WC particle size on high temperature wear resistance of WC-Co cemented carbide. *Hot Work. Technol.* **2022**, *51*, 21–24.
2. Wang, X.L.; Yong, W.; Peng Peng, H. Effect of Abrasive Material on Microstructure and Properties of WC-6% Co Cemented Carbide. *Rare Met. Cem. Carbides* **2022**, *50*, 89–95.
3. Zhou, R.; Zhang, L.H.; He, B.Y.; Liu, Y.H. Numerical simulation of residual stress field in green power metallurgy compacts by modified Drucker–Prager Cap model. *Trans. Nonferrous Met. Soc. China* **2013**, *23*, 2374–2382. [CrossRef]
4. Ding, Q.; Zheng, Y.; Ke, Z.; Zhang, G.T.; Wu, H.; Xu, X.Y.; Lu, X.P.; Zhu, X.G. Effects of fine WC particle size on the microstructure and mechanical properties of WC-8Co cemented carbides with dual-scale and dual-morphology WC grains. *Int. J. Refract. Met. Hard Mater.* **2020**, *87*, 105166. [CrossRef]
5. He, R.; Li, B.; Ou, P.; Yang, C.H.; Yang, H.; Ruan, J.M. Effects of ultrafine WC on the densification behavior and microstructural evolution of coarse-grained WC-5Co cemented carbides. *Ceram. Int.* **2020**, *46*, 12852–12860. [CrossRef]
6. Zhou, H.S.; Lu, K.Z.; He, J.H.; Yang, H.H.; Liu, C.Z. Research progress of ultrasonic compaction powder forming technology. *Tech. Acoust.* **2015**, *34*, 35–42.
7. Wang, X.; Qi, Z.; Chen, W. Study on constitutive behavior of Ti-45Nb alloy under transversal ultrasonic vibration-assisted compression. *Arch. Civ. Mech. Eng.* **2021**, *21*, 1–15. [CrossRef]
8. Zuo, M.M. Study on Dynamic Characteristics of Solid Particles under High-Frequency Vibration. Master's Thesis, Yanshan University, Qinhuangdao, China, 2018.
9. AlHazaa, A.; Haneklaus, N.; Almutairi, Z. Impulse Pressure-Assisted Diffusion Bonding (IPADB): Review and Outlook. *Metals* **2021**, *11*, 323. [CrossRef]
10. Sliva, A.; Brazda, R.; Prochazka, A.; Martynkova, G.S.; Cech Barabaszova, K. Study of the optimum arrangement of sphericalparticles in containers having different cross section shapes. *Nanosci. Nanotechnol.* **2019**, *19*, 2717–2722.
11. Zhao, Y.B.; Ma, L.; Liu, B.; Shang, Z.X. Vibration filling density analysis of pure iron powder based on discrete element method. *Powder Metall. Technol.* **2020**, *38*, 429–435.
12. Wang, W.T.; Wang, J.Y.; Duan, N.Q.; Du, W.H. Study on influencing factors of powder vibration dense packing based on discrete element method. *Chin. Ceram.* **2013**, *49*, 42–45.
13. Sedaghat, H.; Xu, W.; Zhang, L. Ultrasonic vibration-assisted metal forming: Constitutive modelling of acoustoplasticity and applications. *J. Mater. Process. Technol.* **2019**, *265*, 122–129. [CrossRef]
14. Liu, B.; Ma, L.; Liu, Q.Z.; Lu, Y. Effect of EDEM based vibration characteristics on the filling density of NdFeB permanent magnetic powder. *China Powder Sci. Technol.* **2017**, *23*, 72–76.
15. Korim, N.S.; Hu, L. Study the densification behavior and cold compaction mechanisms of solid particles-based powder and spongy particles based powder using a multi-particle finite element method. *Mater. Res. Express* **2020**, *7*, 056509. [CrossRef]
16. Wang, W.; Qi, H.; Liu, P.; Zhao, Y.B.; Chang, H. Numerical Simulation of Densification of Cu–Al Mixed Metal Powder during Axial Compaction. *Metals* **2018**, *8*, 537. [CrossRef]
17. Luo, X.L. Study on Dynamic Mechanical Response of Metal Powder under Impact Loading Based on Three-Dimensional Discrete Element Method. Master's Thesis, Ningbo University, Ningbo, China, 2018.
18. Jia, Q.; An, X.Z.; Zhao, H.Y.; Fu, H.Y.; Fu, H.T.; Zhang, H.; Yang, X.H. Compaction and solid-state sintering of tungsten powders: MPFEM simulation and experimental verification. *J. Alloys Compd.* **2018**, *750*, 341–349. [CrossRef]
19. Ren, X.; Li, Z.; Zheng, Y.; Tian, W.; Zhang, K.C.; Cao, J.R.; Tian, S.Y.; Guo, J.L.; Wen, L.Z.; Liang, G.C. High Volumetric Energy Density of LiFePO4 Battery Based on Ultrasonic Vibration Combined with Thermal Drying Process. *J. Electrochem. Soc.* **2020**, *167*, 130523. [CrossRef]
20. Siegert, K.; Ulmer, J. Influencing the Friction in Metal Forming Processes by Superimposing Ultrasonic Waves. *Cirp. Ann. Manuf. Technol.* **2001**, *50*, 195–200. [CrossRef]
21. Wang, W.T. Research on Numerical Simulation of Iron Powder Molding Based on Discrete Element Method. Ph.D. Thesis, Central North University, Taiyuan, China, 2014.

22. Huang, P.Y. *Principle of Powder Metallurgy M*; Metallurgical Industry Press: Beijing, China, 1982.
23. Bo, H.Y. Meso Simulation Study on the Influence of Interface Friction on Iron Powder Compaction. Master's Thesis, Hefei University of Technology, Hefei, China, 2009.
24. Ghafoor, S.; Li, Y.; Zhao, G.; Li, J.H.; Li, F.Y. Deformation characteristics and formability enhancement during ultrasonic-assisted multi-stage incremental sheet forming. *J. Mater. Res. Technol.* **2022**, *18*, 1038–1054. [CrossRef]
25. Cheng, Z.N. Research on Deformation Mechanism and Microstructure Evolution of Ultrasonic Assisted Incremental Forming Materials. Master's Thesis, Shandong University, Jinan, China, 2021.
26. Wang, W.; Xiao, J.; Ran, Z.; Zheng, X.T.; Fu, B. Improvement of density of energetic materials based on ultrasonic assisted isostatic pressing. *Energ. Mater.* **2021**, *29*, 521–529.
27. Chen, C.X. *Quality Control Principle of Cemented Carbide M*; Cemented Carbide Branch of China Tungsten Association: Zhuzhou, China, 2007.

MDPI
St. Alban-Anlage 66
4052 Basel
Switzerland
www.mdpi.com

Materials Editorial Office
E-mail: materials@mdpi.com
www.mdpi.com/journal/materials